T0275699

Modeling and Simulation of Reactive Flows

Modeling and Simulation of Reactive Flows

Álvaro L. De Bortoli
Greice S. L. Andreis
Felipe N. Pereira

ELSEVIER

AMSTERDAM • BOSTON • HEIDELBERG • LONDON • NEW YORK • OXFORD
PARIS • SAN DIEGO • SAN FRANCISCO • SINGAPORE • SYDNEY • TOKYO

Elsevier
Radarweg 29, PO Box 211, 1000 AE Amsterdam, Netherlands
The Boulevard, Langford Lane, Kidlington, Oxford OX5 1GB, UK
225 Wyman Street, Waltham, MA 02451, USA

Library of Congress Cataloging-in-Publication Data
A catalog record for this book is available from the Library of Congress

British Library Cataloguing-in-Publication Data
A catalogue record for this book is available from the British Library

ISBN: 978-0-12-802974-9

For information on all Elsevier publications
visit our website at http://store.elsevier.com/

Typeset by SPi Global, India

Printed and bound in the USA

Acquisition Editor: Lisa Reading
Editorial Project Manager: Natasha Welford
Production Project Manager: Lisa Jones
Designer: Victoria Pearson

Contents

Preface

Numerical methods have evolved in recent decades, more intensely from the 1980s. However, this development cannot be compared with the development that occurred with computers. Virtually every 3 years, a new computer becomes obsolete.

While the solution of incompressible flows has been more frequent, both numerically and analytically, the compressible flow solution is usually obtained through numerical methods. The compressibility adds nonlinearities to the system equations, which makes it hard to obtain analytical solutions. In this context, the solution of reactive streams becomes even more complex.

Reactive flows are complex, both at low or high temperature, because the formulation typically adds to the Navier-Stokes equations a significant number of nonlinear equations due to reactions.

The combustion of hydrogen, for example, includes about 20 elementary chemical reactions and 8 species. So, eight equations, one for each species, would be added to the equations of nonreactive flow. Even for such a simple mechanism, the numerical solution is complex.

For methane combustion, one has about 300 elementary reactions among some tens of chemical species. Biofuels such as methanol and ethanol involve a similar number of elementary reactions as for methane. Complex fuels such as n-heptane and iso-octane involve hundreds of chemical species and thousands of elementary chemical reactions. For diesel and biodiesel there are thousands of chemical species and tens of thousands of elementary reactions.

Reactions that occur in aqueous media involve numerous minerals in the subsoil, about 4000, and tens of solutes. Of these, about 30 minerals and 15 solutes are the most important. Because the reactions in aqueous media are much faster than those occurring with the minerals, aqueous reactions are considered to be in equilibrium (occurring faster) in the subsoil.

Simplifications of chemical kinetics generally become an alternative. Small mechanisms of a low number of species are often reduced using the assumptions of steady-state and partial equilibrium. Large mechanisms are reduced using a combination of techniques such as direct relation graph (DRG), to obtain a skeleton mechanism and techniques based on the sensitivity analysis of the eigenvalues and eigenvectors of the Jacobian matrix of the chemical system to obtain a reduced mechanism.

Thus, reactive flow is complex and compounded by the set of equations of flow and chemical kinetics, which are solved by numerical methods frequently of semi-implicit type.

This book contains seven chapters and two appendices that were organized sequentially. However, readers, based on their experience, can read each chapter independently.

Chapter 1 deals with the chemical equilibrium, both in aqueous solution and in gaseous phase. Chapter 2 discusses chemical kinetics, starting with a description of the reaction rates. Based on steady-state assumptions and partial equilibrium, some reduced kinetic mechanisms are obtained. In the next chapter are deducted equations for reactive flows based on the balance (conservation) of the properties in the control volume. In Chapter 4, a formulation for mixing fluids and the turbulence models based on characteristics of the flow scales are discussed. The Reynolds and Favre averages are discussed. In Chapter 5, models for reactive flows are presented. Initially, techniques for obtaining reduced kinetic mechanisms, such as DRG, sensitivity analysis, ILDM, REDIM, and flamelet are presented. Then models for premixed flames, diffusion flames, and reactive flows in porous media are shown. In Chapter 6, some of principal methods used for the solution of reactive and nonreactive flows are introduced. Also noteworthy are obtaining the generalized coordinates and the application of the boundary conditions. The formulation at low Mach, very useful in the solution of reactive flows, and some techniques for the acceleration of convergence are presented. Finally, Chapter 7 discusses some solutions to diffusion flames, the flow in porous media and the premixed combustion in porous media.

During the preparation of this book, we tried to use relatively simple ways to model complex situations. Understanding the essence of a physical situation may lead researchers to improve the technique, which then will take them to a more detailed analysis.

The topics are described in a basic and objective way. Among the many existing techniques, those that are more direct and frequently used are discussed. In summary, this book aims to share with the readers some experiences gained by the authors in the solution of reactive flows. It is hoped that the readers, relatively quickly, can gain knowledge that can assist them in the modeling and simulation of reactive flows of technical interest.

Acknowledgments

This book refers to research being developed in the group Modeling and Simulation in Reactive Fluid Dynamics at UFRGS-Brazil, coordinated by Prof. Alvaro Luiz de Bortoli. Prof. De Bortoli is grateful to the collaborators from Brazil and other countries, and gratefully acknowledges the sponsor CNPq, National Counsel of Technological and Scientific Development, under the process 304798/2012-6. He also gratefully acknowledges the sponsor of CAPES, Coordination for the Improvement of Higher Education Personnel, under the process 007515/2011-70.

List of Symbols

a_i	Activity of a species
A	Area
B_j	Sensitivity matrix
c_D	Constant of Prandtl model
c_P	Specific heat at constant pressure
C_s	Smagorinsky constant
C	Concentration, Chapman-Rubesin parameter
d, D	Material derivative
D_i	Mass diffusivity, thermal diffusivity
$\mathrm{d}x$	Infinitesimal element in x-direction
$\mathrm{d}y$	Infinitesimal element in y-direction
$\mathrm{d}z$	Infinitesimal element in z-direction
$\mathrm{d}V$	Control volume
e	Specific energy
e^-	Electron
e_{int}	Internal energy
E	Energy, error
Ea	Activation energy
Eh	Electric potential
$\mathrm{Eo}(V)$	Reduction potential
f_i	Surface force
f, F	Functions
F	Faraday's constant
$\vec{F}$	Vector force
$\vec{g}, g_i$	Gravitational acceleration
G	Gibbs free energy, flame front position
h	Specific enthalpy, time-step
H	Enthalpy
I	Ionic strength, identity matrix
I_{AB}	Index of importance
j_i	Diffusive flux
$\dot{m}_i$	Mass flow
n	Number of moles, number of species, exponent of temperature
$\vec{n}$	Normal vector
N	Number of nodes, number of time-steps

$O()$	Order of ()
p	Pressure
pe	Electrochemical potential
pH	Potential of hydrogen
P	Product, probability
$\dot{q}_j$	Heat flow by conduction
$\dot{q}_v$	Volumetric source of heat (internal, chemical)
q_r	Heat transfer due to radiation
Q	Heat of combustion
Q_e	Activity product
$\dot{Q}$	Potential energy
r	Radius
R	Gas constant
$\vec{R}$	Residuum vector
$\vec{S}$	Surface vector
$\vec{S}_i$	Source term
s_L	Laminar flame velocity
s_T	Turbulent flame velocity
S	Entropy, area, stiffness measure
t	Time
T	Temperature, period of time
U_c	Axial velocity
$v_j, \vec{v}$	Velocity vector
(v_x, v_y, v_z)	Velocity vector
V	Volume
$x_j, (x, y, z)$	Cartesian coordinate system
X_i	Molar fraction of a species
y_0	Distance from the wall
Y_i	Mass fraction
w	Vorticity, velocity component in z-direction
$\dot{w}$	Reaction rate
$\vec{W}$	Vector of flow variables
$\dot{W}$	Rate of work crossing the boundaries
W_i	Molecular weight
z_i	Ionic charge of a species
Z	Mixture fraction

SPECIAL SYMBOLS

α	Thermal diffusivity, angle, coefficient
β	Coefficient of thermal expansion, coefficient
$\delta_{i,j}$	Kronecker delta
Δ	Variation, Laplacian, filter size

ϵ	Viscous dissipation, error
η	Kolmogorov length, similarity variable, generalized coordinate
γ	Parameter
γ_i	Activity coefficient
Γ_i	Gamma function
κ	Thermal conductivity, von Kárman constant
λ	Eigenvalue
Λ	Matrix of eigenvalues
μ	Chemical potential, dynamic viscosity, mean
ν', ν_i	Stoichiometric coefficient
ν_T	Turbulent viscosity
ξ	Radio by length relation, generalized coordinate
ρ	Density
σ	Standard deviation
$\sigma_{i,j}$	Stress tensor
τ	Time, tortuosity
$\tau_{i,j}$	Viscous stress tensor
τ_w	Wall shear stress
ϕ, Φ	Variable
Φ	Viscous dissipation
χ	Scalar dissipation rate
ψ	Variable
Ω	Element of volume
∂	Partial derivative
$\vec{\nabla}$	Gradient operator
$\vec{\nabla}.$	Divergence operator
$\vec{\nabla}\times$	Curl operator

SUBSCRIPTS AND SUPERSCRIPTS

SUBSCRIPTS

b	Burned
c	Chemical
cl	Center line
d	Droplet
D	Diffusivity
f	Fluid
F	Fuel
i, j, k	Species, coordinate directions
ig	Ignition
int	Internal

m	Constant
n	Normal, constant
N	Numerical solution
O_2	Oxidizer
r	Radius
ref	Reference
s	Surface
st	Stoichiometric
t, T	Turbulent, true solution
u	Unburned
η	Refers to Kolmogorov
μ	Constant
τ	Refers to friction
0	Reference
$1, 2, \infty$	Refers to free condition

SUPERSCRIPTS

$k, k+1$	Refers to time-step
$n, n+1$	Refers to time-step
i, j, k	Refers to direction
0	Initial
$+, -, ++, \ldots$	Ions charge

List of Abbreviations

BM	Mass transfer number
CM	Conservation of momentum
CV	Control volume
CFL	Courant-Friedrich-Lewy
CHEM	Chemical number
Da	Damköhler number
erf	Error function
Ec	Eckert number
DFS	Depth first search
DNS	Direct numerical simulation
DRG	Direct relation graph
FO	Fourier number
GCI	Grid convergence index
IAP	Ion activity product
ILDM	Intrinsic low dimensional manifold
Ka	Karlovitz number
Le	Lewis number
LES	Large eddy simulation
NOX	Number of oxidation
Nu	Nusselt number
PDF	Probability density function
Pr	Prandtl number
RANS	Reynolds averaged Navier-Stokes
Re	Reynolds number
REDIM	Reaction diffusion manifolds
Sc	Schmidt number
Sh	Sherwood number
SI	Saturation index

Chapter 1

Chemical Equilibrium

Equilibrium models can be used when the reaction is fast or has sufficient time to reach equilibrium [1]. Chemical equilibrium is the state in which the forward and backward reaction rates are equal. In this state, there is no change in the reactants and products concentrations. The thermodynamic equilibrium occurs when a system reaches the chemical, thermal, and mechanical equilibriums. The equilibrium states of a system can be classified as [2]:

- Classical: When a system in equilibrium has no spontaneous tendency to change.
- Partial: When a system is globally balanced, except in some regions.
- Local: When part of the system is in equilibrium.

The state of thermodynamic equilibrium of a single component system can be characterized by two independent variables, for example, the entropy and the volume. For natural processes, it can be written:

$$\mathrm{d}E \leq T\,\mathrm{d}S - p\,\mathrm{d}V, \tag{1.1}$$

where $\mathrm{d}E$ is the energy variation, $\mathrm{d}S$ the entropy change, and $\mathrm{d}V$ the volume change during the process. For a multicomponent system, the composition must also be specified, which can be done by the number of moles of the species i.

To determine the equilibrium state of a chemical reaction, is used the chemical potential, which is defined as:

$$\mu_i = \left(\frac{\partial E}{\partial n_i}\right). \tag{1.2}$$

Thus, the criterion of equilibrium for a multicomponent system with constant pressure and temperature is

$$\sum_{i=1}^{N} \mu_i \mathrm{d}n_i = 0. \tag{1.3}$$

At the reaction equilibrium point, the Gibbs free energy is minimal for constant pressure and temperature. Table 1.1 shows the Gibbs free energy for some common substances found in applications [3].

Modeling and Simulation of Reactive Flows. http://dx.doi.org/10.1016/B978-0-12-802974-9.00001-5

TABLE 1.1 Gibbs Free Energy for Some Substances

Substance	ΔG (kJ/mol)
SiO_2	−856.3
SO_4^{--}	−744.0
H_2CO_3	−623.2
HCO_3^-	−586.8
Ca^{++}	−553.6
CO_3^{--}	−527.0
Al^{+++}	−489.4
Mg^{++}	−455.4
CO_2	−394.4
NaCl	−384.0
k^+	−282.5
Na^+	−261.5
CO	−137.2
Cl^-	−131.2
CH_4	−50.8
C_2H_6	−32.9
C_3H_8	−23.5
O_2	0
H_2	0
N_2	0
C_7H_{16}	8.7
C_8H_{18}	17.3
OH^-	34.3
$C_{11}H_{24}$	43.0
NO_2	51.3
C_2H_4	68.1
NO	86.6
H^+	203.3
C_2H_2	209.2
O	231.8

In what follows, we will discuss the chemical equilibrium in the aqueous phase, common in geochemistry, and in the gaseous phase, common in combustion problems.

1.1 CHEMICAL EQUILIBRIUM IN AQUEOUS SOLUTION

The reaction between water and rocks in geophysical systems is best conceptualized using a model based on the hypothesis of local equilibrium. The reaction rate measured in a lab reflects the dominant reaction mechanism. Some compounds are soluble in water, such as salts containing Li^+, Na^+, k^+, NH_4^+, NO_3^-, $C_2H_3O_2^-$, the majority of clorets, Cl^-, and most sulfides, SO_4^{--}. The ability of a solid to dissolve in water (liquid) depends on the polarity, temperature, and pressure. Aqueous species are usually considered to be in chemical equilibrium in geochemical models, which significantly reduces the number of independent variables [4]. The equilibrium of a reaction is the point at which the Gibbs free energy is minimal for constant temperature and pressure [5]. The chemical potential μ is given by:

$$\mu = \frac{\partial G}{\partial n}, \tag{1.4}$$

where n is the number of moles. The μ depends on temperature, pressure, and the number of moles of each species in the aqueous solution. Consider the reaction:

$$a\,\mathrm{A} + b\,\mathrm{B} = c\,\mathrm{C} + d\,\mathrm{D}. \tag{1.5}$$

The definition of chemical potential results in

$$d\,\mu_\mathrm{D} + c\,\mu_\mathrm{C} - b\,\mu_\mathrm{B} - a\,\mu_\mathrm{A} = 0. \tag{1.6}$$

Aqueous species do not occur in pure form, because their solubility in water is limited [2]. The solubility is the maximum amount of a substance that can be dissolved in a solution [6]. The saturation index indicates the saturation state of a solution with respect to a mineral phase, which is given by

$$\mathrm{SI} = \log\left(\frac{Q}{k_\mathrm{e}}\right), \tag{1.7}$$

where Q is the product of the activities of the dissolved species. When $\mathrm{SI} < 0$, the solution is undersaturated, and when $\mathrm{SI} > 0$, the solution is supersaturated. The solubility of a given solute can be determined with the solubility product. As an example, the solubility reaction of calcite in water is

$$[\mathrm{CaCO_3}] + [\mathrm{H_2O}] = [\mathrm{Ca^{++}}] + [\mathrm{HCO_3^-}] + [\mathrm{OH^-}] \tag{1.8}$$

with constant of equilibrium $k_\mathrm{e} \sim 10^{-12}$. Because $[\mathrm{Ca^{++}}] \sim [\mathrm{OH^-}] \sim [\mathrm{HCO_3^-}] \sim (10^{-12})^{1/3} \sim 10^{-4}$, then the solubility is of 10^{-4} mol/L, which means 0.01 g/L of calcite.

The solubility of solids in liquid generally increases with increasing temperature and the solubility of most gases decreases with increasing temperature. Most reactions have their rates duplicated or triplicated with increasing temperature of 10 °C [3]. Gases such as O_2, N_2 and CO_2 dissolve readily in water.

Thus, the chemical potential of an aqueous species A_i is given by

$$\mu_i = \mu_i^0 + R\,T \ln(a_i). \tag{1.9}$$

In dilute solutions, the activity and molality are numerically equivalent. The molality of the species and the activity are related by

$$a_i = \gamma_i m_i. \tag{1.10}$$

The constant γ_i is the activity coefficient, which considers the reactive decrease in aqueous solution. The activity coefficient approaches unity for dilute solutions. For mineral (pure) the activity takes the unit value, and for gases the activity is replaced by fugacity. At low pressures, the partial pressure and fugacity become numerically equivalent. Table 1.2 [2] shows the partial pressure for some gases in the atmosphere.

The activity of an ion is affected by its surroundings. The electrostatic force varies inversely with the square root of the distance of separation of ions according to Coulomb's law. Thus, the activity coefficients decrease as the concentration of the solution increases.

The equilibrium constant expresses the point of minimum Gibbs free energy for a chemical reaction. The equilibrium criterion becomes:

$$\Delta G^0 = -R\,T[d \ln(a_D) + c \ln(a_C) - b \ln(a_B) - a \ln(a_A)]. \tag{1.11}$$

The left side of the Equation (1.11) represents the Gibbs free energy of reference. Thus, the equilibrium constant defined in terms of Gibbs free energy becomes:

TABLE 1.2 Partial Pressure for Some Gases in the Atmosphere

Gas	Pressure (atm)
N_2	0.78
O_2	0.21
H_2O vapor	0.001–0.23
CO_2	0.0003
CH_4	0.0000015
CO; SO_2	0.000001
H_2; N_2O	0.0000005

$$\ln(k) = \frac{-\Delta G^0}{RT} \tag{1.12}$$

or, according to Eq. (1.11)

$$\ln(k_e) = d\ln(a_D) + c\ln(a_C) - b\ln(a_B) - a\ln(a_A) \tag{1.13}$$

corresponding to

$$k_e = \frac{a_D^d\, a_C^c}{a_B^b\, a_A^a}. \tag{1.14}$$

The principle of electroneutrality requires that the ionic species in the electrolyte solution have equilibrium of charges on a macroscopic scale. The condition of electroneutrality can be expressed by the charge equilibrium between the species in solution, according to

$$\sum_i z_i\, m_i = 0, \tag{1.15}$$

where z_i are the ionic charges of the primary and secondary species. These species in a fluid can donate or accept electrons. For example:

$$e^- + 0.25O_{2(aq)} + H^+ = 0.5H_2O. \tag{1.16}$$

The electrochemical potential, pe, in this case is defined as

$$\text{pe} = -\log(a_e^-) = \frac{-1}{n}\log\left(\frac{Q_e^-}{k_e^-}\right), \tag{1.17}$$

where k_e^- is of order 25.5 at 25 °C, n is the number of consumed/generated electrons, and Q_e^- is the product of the activities of the reaction. The pe = $-\log(a_e^-)$, or activity of the electrons, is an indication that the solution donates electrons and is analogous to the pH, which corresponds to the potential of hydrogen:

$$\text{pH} = -\log[H^+] = -\log(a_{H^+}). \tag{1.18}$$

The pH of a solution is an indication of the tendency of the solution to donate hydrogen ions. In the case of seawater, CO_2 influences the pH according to the reaction

$$H^+ + HCO_3^- = CO_{2(g)} + H_2O. \tag{1.19}$$

In Table 1.3 [2] the main elements found in seawater are shown. Thus, it is seen that the activity coefficient, can be approximated by $2/3$ for ions with charge 1, by $1/5$ for ions with charge 2 and by $1/100$ for ions with charge 3.

Thus, the electric potential of an aqueous solution results from the Nernst equation,

$$\text{Eh} = \frac{-2.303RT}{nF}\log\left(\frac{Q_e^-}{k_e^-}\right) = \frac{2.303R\,T}{F}\text{pe}, \tag{1.20}$$

TABLE 1.3 Main Elements Found in Seawater

Element	Quantity (mg/kg)	Molality	Activity Coefficient
Cl^-	19,350	0.550	0.628
Na^+	10,760	0.475	0.672
SO_4^{--}	2710	0.016	0.169
Mg^{++}	1290	0.040	0.316
Ca^{++}	411	0.006	0.247
k^+	399	0.010	0.628
HCO_3^-	142	0.0015	0.691

where R is the gas constant, T the temperature, and F the constant of Faraday ($96,485\,\mathrm{C\,mol^{-1}}$).

The water can be ionized, $H_2O = H^+ + OH^-$; be oxidized, $H_2O = (1/2)\ O_2 + 2e^- + 2H^+$; or be reduced, $H_2O + e^- = (1/2)\ H_2 + OH^-$.

In the case of oxidation, results for the electrochemical potential

$$\log(a_{e}^{-}) = \mathrm{pe} = 20.78 - \mathrm{pH} + (1/4)\ \log(p_{O_2}) \tag{1.21}$$

and neglecting the term $(1/4)\ \log(p_{O_2})$ (near-surface), we have

$$\log(a_{e}^{-}) = \mathrm{pe} = 20.78 - \mathrm{pH}. \tag{1.22}$$

So, for the electric potential $\mathrm{Eh} \sim 0.059\mathrm{pH}$, results in

$$\mathrm{Eh} = 1.23 - 0.059\mathrm{pH}. \tag{1.23}$$

To calculate the activity coefficients, which provide an indication of how the concentration affects the activity of the species, one can use the Debye-Huckel equation, among others. By Coulomb's law, the activity coefficient decreases as the concentration increases because the electrostatic forces become stronger as the ions approach. Thus, for concentrated solutions, the repulsion effect seems to dominate. Assuming that the ions behave as charged spheres, the Debye-Huckel equation takes the form [2]

$$\log \gamma_i = -\frac{A z_i^2 \sqrt{I}}{1 + a_i^0 B \sqrt{I}}, \tag{1.24}$$

providing the activity coefficient γ_i of an ion with electric charge z_i. In this expression $A \sim 0.509$ and $B \sim 0.3283$ are functions of temperature, a_i^0 is a parameter related to the size of the ion, and I is the ionic strength of the solution, given by

$$I = 0.5 \sum_i m_i \, z_i^2, \tag{1.25}$$

where m_i is the molality of species i and z_i is the electric charge.

The ionic strength can frequently be approximated for an aqueous (sea water) solution as

$$I \sim \frac{1}{2}[m_{Na^+} + 4m_{Ca^{++}} + 4m_{Mg^{++}} + m_{HCO_3^-} + m_{Cl^-} + 4m_{SO_4^{--}}]. \tag{1.26}$$

In the limiting case of dilute solutions, results for the activity coefficient

$$\log(\gamma_i) = -0.5 z_i^2 \sqrt{I}. \tag{1.27}$$

1.2 CHEMICAL EQUILIBRIUM IN THE GASEOUS PHASE

The chemical equilibrium in gaseous phase is valid for reaction rates tending to infinity, which can be used as a reasonable approximation for the combustion of hydrogen. This assumption does not consider the possibility of dissociation of combustion products.

For an ideal gas mixture, the chemical equilibrium [7–9] can be defined as

$$\mu_i = H_i - T\,S_i = \mu_i^0 + R^T \ln\left(\frac{p_i}{p_0}\right) \tag{1.28}$$

with $p_0 = 1$ atm and

$$\mu_i^0 = H_i^0 - T\,S_i^0 + \int_T^{T_{ref}} C_{p_i}\,dT - T\int_T^{T_{ref}} \frac{C_{p_i}}{T}\,dT \tag{1.29}$$

for $i = 1, 2, \ldots, k$. Table 1.4 shows the specific heat at constant pressure for some gases.

TABLE 1.4 Specific Heat at Constant Pressure

Substance	C_p (J/(kg K))
Oxygen	918
Air	1158
Water vapor	4039
Hydrogen	14,304

The equilibrium condition for the reaction j is

$$\sum_{i=1}^{N} \nu_{ij}\mu_i = 0 \tag{1.30}$$

for $j = 1, 2, \ldots, r$.

Defining the equilibrium constant k_{e_j} for the reaction j results in

$$R\,T\,\ln(k_{e_j}) = -\sum_{i=1}^{N} \nu_{ij}\mu_i^0 \tag{1.31}$$

because

$$k_{e_j} = \prod_{i=1}^{k} \left(\frac{p_i}{p_0}\right)^{\nu_{ij}} \tag{1.32}$$

for $j = 1, 2, \ldots, r$.

For the hydrogen-air combustion, $2H_2 + O_2 = 2H_2O$, one obtains for k_e

$$k_e = \frac{p_{H_2O}^2}{p_{H_2}^2\, p_{O_2}} \tag{1.33}$$

and for $N_2 + O_2 = 2NO$

$$k_e = \frac{p_{NO}^2}{p_{N_2}\, p_{O_2}}. \tag{1.34}$$

For an isolated system, with internal energy U, volume V and mass m fixed, the equilibrium condition is given by the variation of entropy

$$(\mathrm{d}S)_{U,V,m} = 0. \tag{1.35}$$

In other systems, the equilibrium is required for given temperature (T), pressure (p), and equivalence ratio ϕ (or mixture fraction). In these cases, the Gibbs free energy replaces the entropy. It is defined by:

$$G = H - T\,S \tag{1.36}$$

and the second law of thermodynamics in terms of the Gibbs free energy (G) is given by

$$(\mathrm{d}G)_{T,P,m} \leq 0, \tag{1.37}$$

which indicates that the Gibbs function always decreases for a spontaneous, isothermal, and isobaric change of a system of fixed mass. In equilibrium,

$$(\mathrm{d}G)_{T,P,m} = 0. \tag{1.38}$$

For an ideal gas mixture, the Gibbs function becomes

$$G_i = G_i^0 + R_u\,T\,\ln\left(\frac{p_i}{p_0}\right) \tag{1.39}$$

and in a reactive system

$$G_{f_i}(T) = G_i^0(T) - \sum_{j=1}^{M} \nu_j' G_j^0(T), \tag{1.40}$$

where $G_{f_i}(T)$ corresponds to the Gibbs free energy of formation and ν_j' to the stoichiometric coefficients of the reactions.

For a mixture of gases, the Gibbs function results in

$$G_{\text{mix}} = \sum_{i=1}^{N} n_i\, G_i(T) = \sum_{i=1}^{N} n_i \left[R_u\, T\, \ln\left(\frac{p_i}{p_0}\right)\right], \tag{1.41}$$

so that in the equilibrium condition $\mathrm{d}G_{\text{mix}} = 0$.

Because k_e corresponds to the ratio between the partial pressures of the products by the partial pressures of the reactants raised their stoichiometric coefficients, for a general system, the Gibbs function results in

$$\Delta G_T^0 = -R_u\, T\, \ln(k_e) \tag{1.42}$$

and

$$k_e = \exp^{-[\Delta G_T^0/(R\,T)]}. \tag{1.43}$$

Then, if $\Delta G_T^0 > 0$ the results are $\ln(k_e) < 0$, $k_e < 1$, which favors the formation of the reactants. If $\Delta G_T^0 < 0$ the results are $\ln(k_e) > 0$, $k_e > 1$, which favors the formation of the products. We can write the Gibbs function in terms of enthalpy and entropy as

$$\Delta G_T^0 = \Delta H^0 - T \Delta S^0; \tag{1.44}$$

then k_e can be written as

$$k_e = \exp^{-\Delta H^0/(R_u\, T)} \exp^{-\Delta S^0/R}. \tag{1.45}$$

Generally, for the energy balance, it is not necessary to include minor species, whose concentration is less than 0.1 percent. However, for the calculation of certain pollutants whose concentrations are small, smaller species should also be included in the calculation.

The species that appear in larger amounts in combustion are H_2O, CO_2, CO, H_2, and N_2, but the nitrogen participates little in the reactions. For fuel-lean mixtures, or mixtures with an excess of O_2, complete conversion of C to CO_2 and of H to H_2O can be assumed. Considering the reaction

$$CO_2 + H_2 = CO + H_2O \tag{1.46}$$

the equilibrium constant can be written in terms of partial pressures, as shown in Table 1.2, as

$$k_e = \frac{p_{CO}\, p_{H_2O}}{p_{CO_2}\, p_{H_2}}. \tag{1.47}$$

As k_e increases with increasing temperature, more CO and H_2O are produced at high temperatures. In this situation, which tends to occur in stoichiometric condition, dissociation of H_2O and CO_2 occur, and the equilibrium condition for these reactions becomes

$$CO_2 = CO + 0.5O_2 \tag{1.48}$$

with

$$k_e = \frac{p_{CO}\sqrt{p_{O_2}}}{p_{CO_2}}, \tag{1.49}$$

and

$$H_2O = H_2 + 0.5O_2 \tag{1.50}$$

with

$$k_e = \frac{p_{H_2}\sqrt{p_{O_2}}}{p_{H_2O}}. \tag{1.51}$$

The dissociation of principal products of combustion, H_2O, CO_2, CO, H_2, and N_2, and the reactions among the species dissociated gives the following species: OH, CO, H, O, N, and NO, among others.

REFERENCES

[1] Kehew A. Applied chemical hydrogeology. New York: Prentice Hall; 2001.
[2] Bethke CM. Geochemical and biogeochemical reaction modeling. UK: Cambridge University Press; 1996.
[3] Krauskopf KB. Introduction to geochemistry. New York: McGraw-Hill; 1967.
[4] Steefel CI, Lasaga AC. Coupled model for transport of multiple chemical species and kinetic precipitation/dissolution reactions with application to reactive flow in single phase hydrothermal systems. Am J Sci 1994;294:592.
[5] Lasaga AC. Chemical kinetics of water-rock interactions. J Geophys Res 1984;89:4009-25.
[6] Giles MR. Diagenesis: a quantitative perspective. Dordrecht: Kluwer Academic Publishers; 1997.
[7] Peters N. Fifteen lectures on laminar and turbulent combustion. Aachen, Germany: Ercoftac Summer School; 1992. Consulted in July 15, 2008, http://www.itv.rwth-aachen.de/fileadmin/LehreSeminar/Combustion/SummerSchool.pdf.
[8] Warnatz J, Maas U, Dibble RW. Combustion physical and chemical fundamentals, modeling and simulation, experiments, pollutant formation. Berlin/Heidelberg: Springer-Verlag; 2006.
[9] Law CK. Combustion physics. New York: Cambridge University Press; 2006.

Chapter 2

Chemical Kinetics

Chemical kinetics studies the rate of chemical reactions and factors that influence them. A chemical process can be divided into a sequence of one or more elementary reactions that typically involve collisions between two molecules (bimolecular reaction) or dissociation/isomerization of a reagent molecule (unimolecular reaction). A trimolecular reaction, which is the collision of three reactant molecules, occurs less frequently. Usually, the mechanisms are described in a single global step, but in reality they occur in a series of elementary steps [1].

The elementary chemical reactions involve a transition between two atomic or molecular states, and these states are separated by activation energy, which determines the rate at which reactions occur. If activation energy is low, the reaction is rapid; otherwise, the reaction will be slower. The time scale on which the chemical reactions occur spans many orders of magnitude.

2.1 REACTION RATES

The reaction rate is the rate at which reactants are consumed, or the rate at which products are formed, and has units of concentration per unit time (e.g., $\mathrm{mol\,dm^{-3}\,s^{-1}}$) [1]. The stoichiometry of the reaction refers to the number of moles of each reactant and the product that appears in the reaction equation. For example, the reaction equation for the complete combustion of hydrogen is [1]:

$$2\mathrm{H_2} + \mathrm{O_2} \rightarrow 2\mathrm{H_2O}, \tag{2.1}$$

where H_2 has stoichiometric coefficient 2, O_2 has coefficient 1, and H_2O has coefficient 2.

The reaction rate may be defined as the rate of change of concentration of a reactant or product divided by its stoichiometric coefficient. For reaction (2.1) the reaction rate is (in $\mathrm{mol\,dm^{-3}\,s^{-1}}$)

$$\dot{w} = -\frac{1}{2}\frac{\mathrm{d}[\mathrm{H_2}]}{\mathrm{d}t} = -\frac{\mathrm{d}[\mathrm{O_2}]}{\mathrm{d}t} = \frac{1}{2}\frac{\mathrm{d}[\mathrm{H_2O}]}{\mathrm{d}t}. \tag{2.2}$$

The negative signal appears when the reaction rate is defined using the concentration of a reactant, because the rate of change of a product is positive and

Modeling and Simulation of Reactive Flows. http://dx.doi.org/10.1016/B978-0-12-802974-9.00002-7

the resultant reaction rate should be positive. The reaction rate can be expressed as being proportional to the concentrations of the reactants, each raised to a power. For the reaction

$$a_1\mathrm{A}_1 + a_2\mathrm{A}_2 + a_3\mathrm{A}_3 + \cdots \overset{k}{\rightarrow} b_1\mathrm{B}_1 + b_2\mathrm{B}_2 + b_3\mathrm{B}_3 + \cdots \tag{2.3}$$

one obtains

$$\dot{w} = k[\mathrm{A}_1]^{a_1}[\mathrm{A}_2]^{a_2}[\mathrm{A}_3]^{a_3}\cdots \tag{2.4}$$

The proportionality constant (k) is called the reaction rate coefficient. The exponent of a concentration corresponds to the order of reaction with respect to this reagent, and the sum of the powers is the overall order of the reaction. The order does not necessarily reflect the stoichiometry of the reaction equation, as for example, [2]

$$2\mathrm{NO} + \mathrm{O}_2 \overset{k}{\rightarrow} 2\mathrm{NO}_2, \quad \dot{w} = k[\mathrm{NO}]^2[\mathrm{O}_2], \tag{2.5}$$

$$\mathrm{CO} + \mathrm{Cl}_2 \overset{k}{\rightarrow} \mathrm{COCl}_2, \quad \dot{w} = k[\mathrm{CO}][\mathrm{Cl}_2]^{3/2}. \tag{2.6}$$

Other reactions can have complex reaction rates, which usually show complicated dependency on chemical species. They can also contain more than one coefficient in the reaction rate formula. An example of reaction with complex reaction rate is [2]

$$2\mathrm{N}_2\mathrm{O} \overset{k_1}{\rightarrow} 2\mathrm{N}_2 + \mathrm{O}_2, \quad \dot{w} = \frac{k_1[\mathrm{N}_2\mathrm{O}]}{1 + k_2[\mathrm{O}_2]}, \tag{2.7}$$

where k_1 and k_2 depend strongly on temperature.

The elementary reactions have simple reaction rates in which the order with respect to each reagent reflects the molecularity of the process, as in the examples:

Unimolecular decomposition:

$$\mathrm{A} \overset{k}{\rightarrow} \mathrm{B}, \quad \dot{w} = k[\mathrm{A}],$$

Bimolecular reaction:

$$\mathrm{A} + \mathrm{B} \overset{k}{\rightarrow} \mathrm{C}, \quad \dot{w} = k[\mathrm{A}][\mathrm{B}],$$

$$\mathrm{A} + \mathrm{A} \overset{k}{\rightarrow} \mathrm{C}, \quad \dot{w} = k[\mathrm{A}]^2.$$

The equation for an overall reaction mechanism is simply the result of all the elementary reactions in the mechanism. The reaction rate, the reaction rate coefficient, and the reaction order are determined by experiments, and the orders are generally not equal to the stoichiometric coefficients of the overall reaction equation. The overall reaction rate may contain the concentration of reactants, products, and catalysts, but generally does not contain concentrations of reactive intermediates, which appear only in the rates of the elementary steps.

The units of the reaction rate coefficients depend on the relation $[k] = (\text{time})^{-1}(\text{concentration}\ )^{1-n}$, where n is the order of reaction. Substituting the units in the formula $\dot{w} = k[\text{A}][\text{B}]$, one obtains

$$(\text{mol}\,\text{dm}^{-3}\,\text{s}^{-1}) = [k](\text{mol}\,\text{dm}^{-3})(\text{mol}\,\text{dm}^{-3}). \tag{2.8}$$

Therefore, for this case,

$$[k] = \frac{(\text{mol}\,\text{dm}^{-3}\,\text{s}^{-1})}{(\text{mol}\,\text{dm}^{-3})(\text{mol}\,\text{dm}^{-3})} = \text{mol}^{-1}\,\text{dm}^{3}\,\text{s}^{-1}. \tag{2.9}$$

The order of reaction is determined experimentally or by means of mathematical models. The most important reactions are of zero order, first order, and second order, while third-order reactions are quite rare and reactions of order greater than three are not known. The reaction order is obtained from experimental data and assumptions about the sequence of elementary steps by which the reaction occurs.

The reaction rate is a differential equation that describes the rate of change of the concentration of a reactant or product over time. In many cases, the reaction rate may be analytically integrated. In Table 2.1, $[\text{A}]_0$ and $[\text{B}]_0$ correspond to the initial concentrations of A and B, respectively.

The concept of half-life is applied to first-order reactions, because it is related directly to the reaction rate coefficient. The half-life of the reaction is the time required for the reagent concentration be reduced to half of its initial value. For a reaction of zero order, $t_{1/2} = [\text{A}]_0/(2k)$; for a first order reaction, $t_{1/2} = (\ln 2)/k$; and for a reaction of second order, $\text{A}+\text{A}=\text{C}$, $t_{1/2} = 1/(k[\text{A}]_0)$. Thus, the half-life of a first-order reaction is independent of the reagent initial concentration.

TABLE 2.1 **Differential and Integral Forms for the Reaction Rate**

Reaction	Order	Differential Form	Integrated Form
$\text{A} \rightarrow \text{C}$	0	$-\dfrac{d[\text{A}]}{dt} = k$	$[\text{A}] = [\text{A}]_0 - kt$
$\text{A} \rightarrow \text{C}$	1	$-\dfrac{d[\text{A}]}{dt} = k[\text{A}]$	$\ln[\text{A}] = \ln[\text{A}]_0 - kt$
$\text{A}+\text{B} \rightarrow \text{C}$	2	$-\dfrac{d[\text{A}]}{dt} = k[\text{A}][\text{B}]$	$kt = \dfrac{1}{[\text{B}]_0 - [\text{A}]_0} \ln \dfrac{[\text{B}]_0[\text{A}]}{[\text{A}]_0[\text{B}]}$

The reaction rate shows the relationship between velocities and concentrations, and the coefficients of the reaction rates strongly depend nonlinearly on the temperature. Van't Hoff observed empirically that each 10 °C temperature rise doubles the reaction rate, but experiments have shown that this ratio varies between 2 and 4. Then, Svante Arrhenius, student of Van't Hoff, proposed another empirical equation with better results, the known Arrhenius equation

$$k(T) = A\mathrm{e}^{-E_\mathrm{a}/(RT)}, \tag{2.10}$$

where A is the pre-exponential factor, E_a the activation energy (J/mol or cal/mol), R the gas constant [8.314 J/(mol K) = 1.987 cal/(mol K)], and T (K) the temperature. This equation has been checked empirically by comparing its behavior with temperature for most reaction rate coefficients, within experimental accuracy, for large temperature ranges.

The activation energy is the minimum energy that the reactant molecules must possess before the reaction can occur. In the kinetic theory of gases, the factor $\mathrm{e}^{-E_\mathrm{a}/(RT)}$ gives the fraction of collisions among molecules that together possess this minimum energy E_a. The activation energy is determined experimentally by performing the reaction at several temperatures [2]. After taking the natural logarithm of Eq. (2.10), one obtains

$$\ln k = \ln A - \frac{E_\mathrm{a}}{R}\left(\frac{1}{T}\right). \tag{2.11}$$

The graph of $(1/T)$ versus $(\ln k)$ should yield a straight line whose slope is proportional to the activation energy, as shown in Figure 2.1.

There are similar expressions to the Arrhenius equation. One of such expressions contains the temperature dependence, which results in

$$k(T) = A\,T^n\,\mathrm{e}^{-E_\mathrm{a}/(RT)}. \tag{2.12}$$

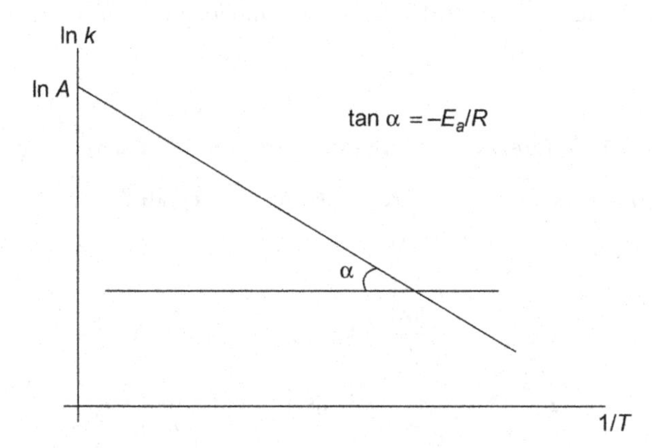

FIGURE 2.1 Line with angular value $(-E_\mathrm{a}/R)$ and coefficient $(\ln A)$.

The reaction rate coefficient may also depend on the pressure. A simple model is the Lindemann model [3]. According to this model, a unimolecular decomposition is only possible if the energy in the molecule is sufficient to break the bond. Therefore, it is necessary that before the decomposition reaction occurs, energy needs to be added to the molecule by collision with other molecules M. So, the excited molecule A^* can be decomposed into products or can disable through a collision,

$$A + M \overset{k_1}{\rightarrow} A^* + M \quad \text{Activation,} \tag{2.13}$$

$$A^* + M \overset{k_2}{\rightarrow} A + M \quad \text{Desactivation,} \tag{2.14}$$

$$A^* \overset{k_3}{\rightarrow} C \quad \text{Unimolecular reaction.} \tag{2.15}$$

The reaction equations are written as

$$\frac{d[C]}{dt} = k_3[A^*], \tag{2.16}$$

$$\frac{d[A^*]}{dt} = k_1[A][M] - k_2[A^*][M] - k_3[A^*]. \tag{2.17}$$

Assuming that the concentration of the reactive intermediate A^* is in steady-state ($d[A^*]/dt = 0$), the concentration of the excited molecule $[A^*]$ and the product C can be defined as

$$[A^*] = \frac{k_1[A][M]}{k_2[M] + k_3}, \quad \frac{d[C]}{dt} = \frac{k_1 k_3[A][M]}{k_2[M] + k_3}, \tag{2.18}$$

where there is a distinction for the reaction at low and high pressure. Under low pressure, the concentration of M is very small. With $k_2[M] \ll k_3$, it follows that

$$\frac{d[C]}{dt} = k_1[A][M] = k_0[A][M], \tag{2.19}$$

where k_0 is the reaction rate coefficient at low pressure. Thus, the reaction rate is proportional to the concentrations of species A and M, because the activation is slow at low pressures.

For high pressure, M has large concentration, and using $k_2[M] \gg k_3$, one obtains

$$\frac{d[C]}{dt} = \frac{k_1 k_3}{k_2}[A] = k_\infty[A], \tag{2.20}$$

where k_∞ is the coefficient of the reaction rate at high pressure. In this case, the reaction rate does not depend on the concentration of M.

In addition to the Lindemann model, there is the theory of unimolecular reactions. If the reaction rate of a unimolecular reaction is written as d[C]/d$t = k$[A], then k depends on the pressure and temperature. For $p \to \infty, k = k_1k_3[\mathrm{M}]/(k_2[\mathrm{M}] + k_3)$ tends to k_∞, that is, k is independent of pressure. For low pressures, k is proportional to [M], resulting in a linear dependence.

In general, the reactions can be classified as homogeneous or heterogeneous. A homogeneous reaction is one that involves a single phase, and a heterogeneous reaction involves more than one phase. The reaction usually occurs at the interface between the two phases or very close to it for heterogeneous reactions.

An irreversible reaction is one that occurs in only one direction and continues in such direction until at least one of the reactants is depleted. An irreversible reaction behaves as if the equilibrium condition does not exist. Actually, no chemical reaction is completely irreversible. A reversible reaction can occur in both directions, depending on the concentrations of reactants and products with regard to equilibrium concentrations. For the reaction

$$a\mathrm{A} + b\mathrm{B} \underset{k_b}{\overset{k_f}{\rightleftharpoons}} c\mathrm{C} + d\mathrm{D} \tag{2.21}$$

the equilibrium concentrations are related by the following thermodynamic relation

$$K_\mathrm{e} = \frac{k_f}{k_b} = \frac{[\mathrm{C}]^c[\mathrm{D}]^d}{[\mathrm{A}]^a[\mathrm{B}]^b}, \tag{2.22}$$

where the subscript f refers to the forward reaction and b to the backward reaction. The equilibrium constant decreases with increasing temperature for exothermic reactions and increases with increased temperature for endothermic reactions.

In general, for a mechanism with a single reversible reaction involving n species, the reaction rate is given by [4]

$$\dot{w} = \dot{w}_f - \dot{w}_b = k_f \prod_{i=1}^{n} [\mathrm{A}_i]^{\nu_{fi}} - k_b \prod_{i=1}^{n} [\mathrm{A}_i]^{\nu_{bi}} \tag{2.23}$$

and for a mechanism with j steps [5],

$$\dot{w}_j = \dot{w}_{fj} - \dot{w}_{bj} = k_{fj} \prod_{i=1}^{n} [\mathrm{A}_i]^{\nu_{fij}} - k_{bj} \prod_{i=1}^{n} [\mathrm{A}_i]^{\nu_{bij}}, \tag{2.24}$$

or in terms of mass fractions,

$$\dot{w}_j = k_{fj} \prod_{i=1}^{n} \left(\frac{\rho Y_i}{W_i}\right)^{\nu_{fij}} - k_{bj} \prod_{i=1}^{n} \left(\frac{\rho Y_i}{W_i}\right)^{\nu_{bij}}, \tag{2.25}$$

where ρ is the density and W_i the molecular weight of species i.

The determination of concentrations of reactants and products as a function of time is a basic procedure of chemical kinetics. There are methods to evaluate changes in the concentration of reactants and products, and the choice of the appropriate experimental technique depends on the reaction rate. The following methods are generally employed:

1. Ultraviolet and visible spectroscopy
2. Mass spectrometry
3. Gas chromatography
4. Nuclear magnetic resonance
5. Electron spin resonance
6. Potentiometry

2.2 KINETIC MECHANISMS

In the 1970s, researchers analyzed flames of simple fuels such as hydrogen, carbon monoxide, methane, and methanol [6]. In the oxidation mechanism of hydrogen (H_2), for example, eight species and 19 elementary reactions are frequently used, and in the oxidation of iso-octane (iC_8H_{18}) over 3600 reversible reactions appear among about 860 chemical species [7].

The kinetic mechanisms can contain basically three types of reactions: in series, in parallel, and independent [2]. In parallel reactions (competitive reactions), the reagent is consumed by two different reaction paths to form different products:

$$A \overset{k_1}{\rightarrow} B, \tag{2.26}$$

$$A \overset{k_2}{\rightarrow} C. \tag{2.27}$$

In reactions occurring in series (consecutive reactions), reactant forms an intermediate product, which subsequently reacts to form another product:

$$A \overset{k_1}{\rightarrow} B \overset{k_2}{\rightarrow} C. \tag{2.28}$$

The kinetic mechanisms involve a combination of both reactions in series or parallel, such as in the formation of butadiene C_4H_6 from ethanol:

$$C_2H_5OH \rightarrow C_2H_4 + H_2O, \tag{2.29}$$

$$C_2H_5OH \rightarrow CH_3CHO + H_2, \tag{2.30}$$

$$C_2H_4 + CH_3CHO \rightarrow C_4H_6 + H_2O. \tag{2.31}$$

The chain reactions have complex kinetics and occur quickly, and these reactions form the basis of combustion processes. In a chain reaction, a reactive intermediate reacts to produce another intermediate, and so on. The intermediate species in a chain reaction is a chain propagator, which in many cases is a radical.

The first step in a chain reaction is the initiation, which is the formation of the propagating chain from a reagent. Afterwards, propagation occurs, where the propagator reacts with one molecule of the reactant to form other propagator. Radicals combine themselves, and the chain ends in a process called termination.

Because the chain reactions occur with high speed, many fast reactions happen. The formation of a single reactive intermediate originates many products before ending the chain. Thus, the rate of formation of products is much higher than the speed of the initiation step.

Consider the most important reactions for the ignition of hydrogen, shown in Table 2.2. The mechanism consists of the following steps:

1. Initiation of the chain: Reactive species (radicals characterized by "$*$") are formed from stable species (reaction 1).
2. Propagation of the chain: Intermediate species react with stable species to form other reactive species (reaction 2 or 6).
3. Branch of the chain: intermediate species react with stable species to form two reactive species (reactions 3 and 4).
4. Termination of the chain: intermediate species react to form stable species (reaction 5).

Adding the chain initiation and propagation reactions, the radicals of the mechanism are formed by the reactants H_2 and O_2 ($2H_2 + O_2 = H^* + OH^* + H_2O$).

TABLE 2.2 Important Reactions for the Hydrogen Ignition Mechanism [3]

N	Reaction	Step
1	$H_2 + O_2 = 2OH^*$	Initiation of the chain
2	$OH^* + H_2 = H_2O + H^*$	Propagation of the chain
3	$H^* + O_2 = OH^* + O^*$	Branch of the chain
4	$O^* + H_2 = OH^* + H^*$	Branch of the chain
5	$H^* = 1/2H_2$	Termination of the chain
6	$H^* + O_2 + M = HO_2^* + M$	Propagation of the chain

2.2.1 Reactions of Oxidation Reduction

An oxidation-reduction reaction (redox) corresponds to a chemical reaction that involves the transference of electrons from one species to another. In redox reactions, as the proper name suggests, there are simultaneous occurrence of the phenomena of oxidation and reduction, because the existence of one depends on the occurrence of the other.

Oxidation corresponds to a chemical process that results from the removal of electrons from one species, with an increase of its oxidation number (NOX). The reduction is a process in which a species receives electrons, resulting in a reduction of NOX.

The oxidizer is a compound that causes the oxidation of other substances, receiving electrons. Thus, the oxidizer is reduced during the redox reaction. The reducer causes the reduction of other substances transferring electrons to them. The reducer undergoes oxidation during the redox reaction.

Strong reducers correspond to elements with low electronegativity or species with low ionization energy. The alkali and alkaline earth metals are strong reducers, such as Li and Na.

Strong oxidizers are generally elements with high electronegativity, molecules or ions containing these elements, and some metals of high NOX. The most common oxidizers are halogens, such as F_2 and Cl_2; oxyacid and oxyanions, such as NO_3^-, IO_3^-, and MnO_4^-; forms of oxygen, such as O_3; and peroxides, such as H_2O_2.

A particular case of redox reactions is known as self-redox, in which the same element undergoes oxidation and reduction in the same reaction. This happens between halogens and alkali hydroxides. An example is the reaction of sodium hydroxide with chlorine to form chloride and sodium hypochlorite:

$$2\,NaOH + Cl_2 \rightarrow NaCl + NaClO + H_2O. \tag{2.32}$$

In this reaction the chlorine (NOX $= 0$) is oxidized to hypochlorite (NOX $= +1$) and is also reduced to chloride (NOX $= -1$).

Oxidation Number (NOX)

The oxidation number of a chemical element refers to the number of charges that an atom would possess if the electrons were not shared but located entirely on a single atom.

For example, in the molecule H_2O, there is a covalent bond among the hydrogen atoms and the oxygen atom. These electrons are shared between oxygen and hydrogen. When the electrons are uniquely assigned to one of the atoms, because oxygen is more electronegative than the hydrogen, the representation of the molecule becomes

$$H^+ - O^{-2} - H^+, \tag{2.33}$$

that is, the oxidation number of each hydrogen atom is +1, while for the oxygen it is −2.

For the determination of the NOX of a molecule, there are some basic rules:

1. The sum of the oxidation numbers of all atoms that constitute a molecule is equal to zero (electroneutrality rule).
2. The sum of the oxidation numbers of all atoms of a polyatomic ion is always equal to the charge of this ion.
3. Isolated elements and simple substances have NOX = 0.
4. Monatomic ions have the oxidation number equal to its own charge.
5. Alkali metals (Li, Na, K, Rb, Cs, Fr) and silver (Ag) have NOX = +1.
6. Alkaline earth metals (Be, Mg, Ca, Sr, Ba, Ra) and zinc (Zn) have NOX = +2.
7. In compounds, aluminum (Al) has NOX = +3.
8. In compounds, hydrogen (H) has NOX = +1, except for metal hydrides, where it has NOX = −1.
9. Oxygen (O_2) has NOX = −2, except for peroxides, where the NOX = −1, and superoxide, where NOX = −0.5.
10. In the right side of the chemical formula, the halogens (F, Cl, Br, and I) have NOX = −1.

For example, in the molecule H_2SO_4, each H has NOX = +1 (rule 8), each O has NOX = −2 (rule 9), and, in order to maintain the charge neutrality of the molecule, S has NOX = +6 (rule 1).

Concepts of Half-Reaction

The redox reactions can be expressed as a combination of two partial ionic reactions: an oxidation half-reaction and a reduction half-reaction. In general, the half-reactions of oxidation and reduction are written as follows, respectively:

$$\mathrm{Red_1} \rightarrow \mathrm{Ox_1} + \upsilon e^-, \tag{2.34}$$

$$\mathrm{Ox_2} + \upsilon e^- \rightarrow \mathrm{Red_2}. \tag{2.35}$$

For example, in the reaction in which two hydrogen molecules react with one molecule of oxygen, they form two molecules of water

$$2H_2 + O_2 \rightarrow 2H_2O. \tag{2.36}$$

In this reaction, hydrogen is oxidized, going from NOX = 0 to NOX = +1, and oxygen is reduced, going from NOX = 0 to NOX = −2. The half-reactions of oxidation and reduction of these compounds correspond to the following expressions, respectively,

$$H_2 \rightarrow 2H^+ + 2e^-, \tag{2.37}$$

$$O_2 + 4e^- + 2H^+ \rightarrow 2OH^-. \tag{2.38}$$

When performing a linear combination of the two equations, the number of electrons donated by the reducer is equal to the number of electrons received by the oxidizer, giving

$$\begin{array}{l} + \left\{ \begin{array}{rcl} 2H_2 & \rightarrow & 4H^+ + 4e^- \\ O_2 + 4e + 2H^+ & \rightarrow & 2OH^- \end{array} \right. \\ \hline 2H_2 + O_2 \rightarrow 2H^+ + 2OH^- \end{array}. \tag{2.39}$$

Because the ions H^+ and OH^- unite to form H_2O, one obtains

$$2H_2 + O_2 \rightarrow 2H_2O, \tag{2.40}$$

which corresponds to the overall reaction.

Potential of Reduction

The reduction potential indicates the tendency of a chemical species to undergo reduction. The higher the reduction potential, the greater the compound affinity for electrons and the greater its tendency to be reduced.

The value of an electrode potential is given in relation to a reference electrode. It is considered a standard hydrogen electrode $\left[Pt_{(s)}|H_{2(g)}|H^+_{(aq)}\right]$ with H^+ ion activity equal to 1 and fugacity of the gas H_2 equal to 1 bar. The potential of the hydrogen electrode is made equal to zero, and the other potentials are written in relation to this value [8].

Because the measurement of the hydrogen reduction potential in laboratories is difficult to implement, electrodes of Ag/AgCl are often used as reference. Table 2.3 [9] shows the potential of the main electrodes, where Eo(V) represents the reduction potential given in volts.

The next section describes some procedures for obtaining reduced kinetic mechanisms.

2.3 REDUCED KINETIC MECHANISMS

Simulations with detailed kinetic mechanisms are complicated by the existence of highly reactive radicals, producing significant stiffness in the system of equations. This is due to differences in the time scales of the reactions. Thus, there is a need to develop reduced mechanisms with fewer variables and moderate stiffness, while maintaining the accuracy and completeness of detailed kinetic mechanisms [10].

The use of detailed kinetic mechanisms implies the solution of hundreds of conservation equations for hydrocarbons. Therefore, it is convenient to use simplified kinetic mechanisms that describe the reaction system in terms of few species. This can be done using reduced kinetic mechanisms based on assumptions of steady-state and partial equilibrium [3]. These reduced mechanisms are

TABLE 2.3 Potential of Some Electrodes

Half-reaction	Eo(V)
$Li^{+}_{(aq)} + e^{-} \rightarrow Li_{(s)}$	−3.05
$K^{+}_{(aq)} + e^{-} \rightarrow K_{(s)}$	−2.93
$Ca^{2+}_{(aq)} + 2\,e^{-} \rightarrow Ca_{(s)}$	−2.87
$Na^{+}_{(aq)} + e^{-} \rightarrow Na_{(s)}$	−2.71
$Mg^{2+}_{(aq)} + 2\,e^{-} \rightarrow Mg_{(s)}$	−2.37
$Al^{3+}_{(aq)} + 3\,e^{-} \rightarrow Al_{(s)}$	−1.66
$2\,H_2O + 2\,e^{-} \rightarrow H_{2\,(g)} + 2\,OH^{-}_{(aq)}$	−0.83
$Zn^{2+}_{(aq)} + 2\,e^{-} \rightarrow Zn_{(s)}$	−0.76
$Cr^{3+}_{(aq)} + 3\,e^{-} \rightarrow Cr_{(s)}$	−0.74
$Fe^{2+}_{(aq)} + 2\,e^{-} \rightarrow Fe_{(s)}$	−0.44
$PbSO_{4\,(s)} + 2\,e^{-} \rightarrow Pb_{(s)} + SO_{2\,(aq)}^{-4}$	−0.31
$Ni^{2+}_{(aq)} + 2\,e^{-} \rightarrow Ni_{(s)}$	−0.25
$Sn^{2+}_{(aq)} + 2\,e^{-} \rightarrow Sn_{(s)}$	−0.14
$Pb^{2+}_{(aq)} + 2\,e^{-} \rightarrow Pb(s)$	−0.13
$2\,H^{+}_{(aq)} + 2\,e^{-} \rightarrow H_{2\,(g)}$	**0.00**
$Cu^{2+}_{(aq)} + e^{-} \rightarrow Cu^{+}_{(aq)}$	0.15
$AgCl_{(s)} + e^{-} \rightarrow Ag(s) + Cl^{-}_{(aq)}$	0.22
$Cu^{2+}_{(aq)} + 2\,e^{-} \rightarrow Cu(s)$	0.34
$O_{2\,(g)} + 2\,H_2O + 4\,e^{-} \rightarrow 4\,OH^{-}_{(aq)}$	0.40
$Ag^{+}_{(aq)} + e^{-} \rightarrow Ag(s)$	0.80
$O_{2\,(g)} + 4\,H^{+}_{(aq)} + 4\,e^{-} \rightarrow 2\,H_2O$	1.23
$Cl_{2\,(g)} + 2\,e^{-} \rightarrow 2\,Cl^{-}_{(aq)}$	1.36
$Au^{3+}_{(aq)} + 3\,e^{-} \rightarrow Au_{(s)}$	1.50
$F_{2\,(g)} + 2\,e^{-} \rightarrow 2\,F^{-}_{(aq)}$	2.87

only valid under certain conditions, that is, they produce more reliable results for certain temperature ranges. A mechanism for a given fuel may produce good results in the simulation of non-premixed flames but unsatisfactory results in the simulation of premixed flames.

In the 1980s reduced mechanisms for premixed and non-premixed flames of methane were described and, shortly after, these mechanisms were used in the asymptotic and numerical analysis [11]. In the 1980s, some research groups focused their attention on methane flames and developed useful techniques for the systematic reduction of the detailed kinetic mechanism. It was found that kinetic models for hydrocarbons have a logical hierarchical structure, where the kinetic mechanism of any fuel contains, as a subset, the mechanism of smaller molecules [12].

Reduced mechanisms are obtained by introducing appropriate assumptions of steady-state and partial equilibrium in the detailed mechanism and neglecting the terms and reactions of less importance, obtaining a simplified description of the flame structure [13]. This methodology has great potential for the hydrocarbon flames, because the chemical kinetics of flames for most of these compounds originates from chain reactions, where each intermediate species is produced and consumed by just some main reactions.

In a homogeneous system, the assumption of steady-state is valid for the intermediate species that are produced by slow reactions and consumed by fast reactions, so that their concentrations remain small.

Consider the mechanism

$$\mathrm{A} \overset{k_1}{\to} \mathrm{B} \overset{k_2}{\to} \mathrm{C}, \tag{2.41}$$

whose reaction rates are given by:

$$\frac{\mathrm{d}[\mathrm{A}]}{\mathrm{d}t} = -k_1[\mathrm{A}], \tag{2.42}$$

$$\frac{\mathrm{d}[\mathrm{B}]}{\mathrm{d}t} = k_1[\mathrm{A}] - k_2[\mathrm{B}], \tag{2.43}$$

$$\frac{\mathrm{d}[\mathrm{C}]}{\mathrm{d}t} = k_2[\mathrm{B}]. \tag{2.44}$$

For $k_2 \gg k_1$, the species B can be placed in steady-state, that is,

$$\frac{\mathrm{d}[\mathrm{B}]}{\mathrm{d}t} = k_1[\mathrm{A}] - k_2[\mathrm{B}] \sim 0 \tag{2.45}$$

and, thus, $[\mathrm{B}] = k_1[\mathrm{A}]/k_2$.

In many engineering problems, it is acceptable assume steady-state for intermediate species with concentrations lower than 1 percent.

The use of partial equilibrium assumption is justified when the coefficients of the reaction rates of the forward and backward reactions are much larger than all the other coefficients of the reaction mechanism [14].

TABLE 2.4 Skeletal Mechanism for Hydrogen (Units are mol, cm^3, s, K, and cal/mol) [15]

N	Reaction	A	n	E_a	k ($T = 1900$ K)
1f	$H + O_2 \rightarrow O + OH$	3.55E+15	−0.406	1.66E + 04	2.01E+12
1b	$O + OH \rightarrow H + O_2$	1.03E+13	−0.015	−1.33E + 02	9.50E+12
2f	$O + H_2 \rightarrow H + OH$	5.08E+04	2.670	6.29E + 03	5.42E+12
2b	$H + OH \rightarrow O + H_2$	2.64E+04	2.651	4.88E + 03	3.55E+12
3f	$OH + H_2 \rightarrow H + H_2O$	2.16E+08	1.510	3.43E + 03	7.75E+12
3b	$H + H_2O \rightarrow OH + H_2$	2.29E+09	1.404	1.83E + 04	7.05E+11
4f	$OH + OH \rightarrow O + H_2O$	1.45E+05	2.107	−2.90E + 03	2.55E+12
4b	$O + H_2O \rightarrow OH + OH$	2.97E+06	2.020	1.34E + 04	3.54E+11
5f	$H + O_2 + M \rightarrow HO_2 + M$	1.48E+12	0.600	0.00E + 00	1.37E+14
5b	$HO_2 + M \rightarrow H + O_2 + M$	3.09E+12	0.528	4.89E + 04	3.80E+08
6f	$HO_2 + H \rightarrow OH + OH$	7.08E+13	0.000	2.95E + 02	6.55E+13
6b	$OH + OH \rightarrow HO_2 + H$	2.03E+10	0.720	3.68E + 04	2.60E+08
7f	$HO_2 + H \rightarrow H_2 + O_2$	1.66E+13	0.000	8.23E + 02	1.33E+13
7b	$H_2 + O_2 \rightarrow HO_2 + H$	3.17E+12	0.348	5.55E + 04	1.71E+07
8f	$HO_2 + OH \rightarrow H_2O + O_2$	1.97E+10	0.962	−3.28E + 02	3.07E+13
8b	$H_2O + O_2 \rightarrow HO_2 + OH$	3.99E+10	1.204	6.93E + 04	3.58E+06

Consider the mechanism of hydrogen presented in Table 2.4 [15]. A numerical or experimental analysis shows that for high temperatures, $T > 1800$ K and $p = 1$ bar, the reaction rate coefficients of forward and backward reactions are so high that the partial equilibrium assumption can be considered for the reactions 1, 2, and 3. In this case, each reaction is in equilibrium and, therefore, the coefficients of the reaction rates of the forward and backward reactions are equal, that is,

$$k_{1f}[H][O_2] = k_{1b}[O][OH], \tag{2.46}$$

$$k_{2f}[O][H_2] = k_{2b}[H][OH], \tag{2.47}$$

$$k_{3f}[OH][H_2] = k_{3b}[H][H_2O], \tag{2.48}$$

from which it follows that

$$[H] = \left(\frac{k_{1f} k_{2f} k_{3f}^2 [O_2][H_2]^3}{k_{1b} k_{2b} k_{3b}^2 [H_2O]^2} \right)^{1/2}, \tag{2.49}$$

$$[\mathrm{O}] = \frac{k_{1f}k_{3f}[\mathrm{O_2}][\mathrm{H_2}]}{k_{1b}k_{3b}[\mathrm{H_2O}]}, \tag{2.50}$$

$$[\mathrm{OH}] = \left(\frac{k_{1f}k_{2f}[\mathrm{O_2}][\mathrm{H_2}]}{k_{1b}k_{2b}}\right)^{1/2}. \tag{2.51}$$

2.3.1 Assumptions of Steady-State and Partial Equilibrium

The assumptions of steady-state and partial equilibrium help to simplify the system of kinetic equations and minimize the computational time required for its solution, but these assumptions are not applied in low temperature regimes. These hypotheses can be used when the error between the simplified and complete model solutions is reduced to an acceptable level [16].

One way to obtain a reduced kinetic mechanism is as follows:

1. Estimate the order of magnitude of the coefficients of reaction rates.
2. Define the main chain.
3. Introduce the hypotheses of partial equilibrium and steady-state.
4. Identify the overall reactions.
5. Determine reaction rates through an asymptotic analysis.

To implement this strategy, consider the hydrogen mechanism given in Table 2.4. Through the modified Arrhenius equation (2.12), with $T = 1900\,\mathrm{K}$, the order of magnitude of the coefficients of reaction rates is estimated and defined a main chain for the process: $H_2 - H_2O$. As noted in the last column of Table 2.4, reactions 1, 2, and 3 can be considered to be in partial equilibrium. The concentrations of H, O, and OH can be calculated using Eqs. (2.49)–(2.51). Applying the steady-state hypothesis for the species HO_2, OH, and O, one obtains the following two-step mechanism for hydrogen:

$$\text{I} \quad 3\mathrm{H_2} + \mathrm{O_2} \rightarrow 2\mathrm{H_2O} + 2\mathrm{H}, \tag{2.52}$$

$$\text{II} \quad \mathrm{II} + \mathrm{H} + \mathrm{M} \rightarrow \mathrm{H_2} + \mathrm{M}. \tag{2.53}$$

To obtain the rates of this mechanism an asymptotic analysis is employed, which consists of assuming the steady-state hypothesis for certain species and obtaining algebraic equations among the reaction rates. The overall mechanism for the species in a transient regime is defined by the stoichiometry of the reactions. Thus, the overall mechanism depends on the choice of reaction rates to be eliminated. The fastest reaction rate by which it is consumed is chosen for each species. For the case of the mechanism presented in Table 2.4, the balance equations for hydrogen mechanism can be written as

$$w_{\mathrm{H_2}} = -w_2 - w_3 + w_7, \tag{2.54}$$

$$w_{\mathrm{O_2}} = -w_1 - w_5 + w_7 + w_8, \tag{2.55}$$

$$w_{H_2O} = w_3 + w_4 + w_8, \tag{2.56}$$
$$w_H = -w_1 + w_2 + w_3 - w_5 - w_6 - w_7, \tag{2.57}$$
$$w_O = w_1 - w_2 + w_4, \tag{2.58}$$
$$w_{OH} = w_1 + w_2 - w_3 - 2w_4 + 2w_6 - w_8, \tag{2.59}$$
$$w_{HO_2} = w_5 - w_6 - w_7 - w_8. \tag{2.60}$$

The positive sign refers to species that appear on the right side of an elementary reaction (products), while the negative sign refers to species on the left side (reactants). For example, in the reaction 1f (Table 2.4): $H + O_2 \rightarrow O + OH$, $w_H = -w_1$, and $w_{OH} = +w_1$, repeating this procedure for all other species and reactions that appear in Table 2.4. Note that:

$$w_1 = w_{1f} - w_{1b}, \tag{2.61}$$
$$w_2 = w_{2f} - w_{2b}, \tag{2.62}$$
$$w_3 = w_{3f} - w_{3b}, \tag{2.63}$$
$$w_4 = w_{4f} - w_{4b}, \tag{2.64}$$
$$w_5 = w_{5f} - w_{5b}, \tag{2.65}$$
$$w_6 = w_{6f} - w_{6b}, \tag{2.66}$$
$$w_7 = w_{7f} - w_{7b}, \tag{2.67}$$
$$w_8 = w_{8f} - w_{8b}. \tag{2.68}$$

It is assumed that the species HO_2, OH, and O are in steady state and, therefore, w_i is zero for these species, leading to three algebraic equations among the reaction rates. Eliminating those rates of faster consumption, that is, w_2 in the equation for O, w_7 for HO_2, and w_8 for OH, results in:

$$w_2 = w_1 + w_4, \tag{2.69}$$
$$w_7 = -2w_1 + w_3 + w_4 + w_5 - 3w_6, \tag{2.70}$$
$$w_8 = 2w_1 - w_3 - w_4 + 2w_6. \tag{2.71}$$

Making the reaction rates w_I and w_{II} equal to

$$w_I = w_1 + w_6, \tag{2.72}$$
$$w_{II} = w_5, \tag{2.73}$$

one obtains the linear combinations:

$$w_{H_2} = -3w_1 + w_5 - 3w_6 = -3w_I + w_{II}, \tag{2.74}$$
$$w_{O_2} = -w_1 - w_6 = -w_I, \tag{2.75}$$
$$w_{H_2O} = 2w_1 + 2w_6 = 2w_I, \tag{2.76}$$
$$w_H = 2w_1 - 2w_5 + 2w_6 = 2w_I - 2w_{II}. \tag{2.77}$$

Therefore, the stoichiometry of the reactions corresponds to the mechanism given in Eqs. (2.52) and (2.53), with reaction rates:

$$w_{\mathrm{I}} = w_1 + w_6 = k_{1f}[\mathrm{H}][\mathrm{O_2}] - k_{1b}[\mathrm{O}][\mathrm{OH}] + k_{6f}[\mathrm{HO_2}][\mathrm{H}] - k_{6b}[\mathrm{OH}]^2, \tag{2.78}$$

$$w_{\mathrm{II}} = w_5 = k_{5f}[\mathrm{H}][\mathrm{O_2}][\mathrm{M}] - k_{5b}[\mathrm{HO_2}][\mathrm{M}]. \tag{2.79}$$

The concentration of the species HO_2 can be expressed in terms of concentrations of the species known [11]. Making $w_{\mathrm{HO_2}} = 0$ in Eq. (2.60), gives

$$[\mathrm{HO_2}] = \frac{k_{5f}[\mathrm{H}][\mathrm{O_2}][\mathrm{M}] + k_{6b}[\mathrm{OH}]^2 + k_{7b}[\mathrm{H_2}][\mathrm{O_2}] + k_{8b}[\mathrm{H_2O}][\mathrm{O_2}]}{k_{5b}[\mathrm{M}] + k_{6f}[\mathrm{H}] + k_{7f}[\mathrm{H}] + k_{8f}[\mathrm{OH}]}, \tag{2.80}$$

whose value is frequently small, $\sim 10^{-5}\%$ mass, and using the concentrations of species H and OH given in Eqs. (2.49) and (2.51), respectively, one obtains

$$[\mathrm{HO_2}] = \frac{\left\{\left[\left(\frac{k_{1f}k_{2f}k_{3f}^2k_{5f}^2[\mathrm{O_2}][\mathrm{H_2}][\mathrm{M}]^2}{k_{1b}k_{2b}k_{3b}^2[\mathrm{H_2O}]^2}\right)^{1/2} + \frac{k_{1f}k_{2f}k_{6b}}{k_{1b}k_{2b}}\right][\mathrm{H_2}] + k_{8b}[\mathrm{H_2O}]\right\}[\mathrm{O_2}]}{k_{5b}[\mathrm{M}] + \left(\frac{k_{1f}k_{2f}k_{3f}^2(k_{6f}+k_{7f})^2[\mathrm{O_2}][\mathrm{H_2}]^3}{k_{1b}k_{2b}k_{3b}^2[\mathrm{H_2O}]^2}\right)^{1/2} + \left(\frac{k_{1f}k_{2f}k_{8f}^2[\mathrm{O_2}][\mathrm{H_2}]}{k_{1b}k_{2b}}\right)^{1/2}}. \tag{2.81}$$

2.3.2 Reduced Kinetic Mechanisms for Some Fuels

The principal fuels used in combustion processes are combinations of carbon and hydrogen called hydrocarbons. The alcohols are combinations of carbon, hydrogen, and oxygen. Neglecting the presence of nitrogen results in the following reactions for the global mechanism of some important fuels:

1. Hydrogen: $2H_2 + O_2 \rightarrow 2H_2O$
2. Methane: $CH_4 + 2O_2 \rightarrow CO_2 + 2H_2O$
3. Propane: $C_3H_8 + 5O_2 \rightarrow 3CO_2 + 4H_2O$
4. Methanol: $CH_3OH + \frac{3}{2}O_2 \rightarrow CO_2 + 2H_2O$
5. Ethanol: $C_2H_5OH + 3O_2 \rightarrow 2CO_2 + 3H_2O$
6. n-Heptane: $nC_7H_{16} + 11O_2 \rightarrow 7CO_2 + 8H_2O$
7. Iso-octane: $iC_8H_{18} + \frac{25}{2}O_2 \rightarrow 8CO_2 + 9H_2O$
8. Diesel (a model): $C_{16}H_{34} + \frac{49}{2}O_2 \rightarrow 16CO_2 + 17H_2O$
9. Biodiesel (a model): $C_{19}H_{36}O_2 + 27O_2 \rightarrow 19CO_2 + 18H_2O$

A reduced kinetic mechanism has been preferred to maintain the accuracy of the results and reduce the stiffness of the equations of the reactive system. A reduction of the system of equations is obtained by the introduction of appropriate approximations. Frequently, species O and OH are considered to be in steady-state.

Schemes of four to 10 steps usually provide reasonable precision at low cost for fuels of low chain, but depend on the conditions of the flame. Various reduced kinetic mechanisms are found in the literature, and their final form depends on the flame temperature [11, 17]. The reduced kinetic mechanism can be obtained from the analysis described before, starting with a detailed mechanisms found in the literature. Some reduced mechanism are shown as follows:

Hydrogen (H_2): 2 steps [18]

$$3H_2 + O_2 \rightarrow 2H + 2H_2O, \tag{2.82}$$

$$H + H + M \rightarrow H_2 + M. \tag{2.83}$$

Methane (CH_4): 4 steps [5]

$$CH_4 + 2H + H_2O \rightarrow CO + 4H_2, \tag{2.84}$$

$$CO + H_2O \rightarrow CO_2 + H_2, \tag{2.85}$$

$$3H_2 + O_2 \rightarrow 2H + 2H_2O, \tag{2.86}$$

$$H + H + M \rightarrow H_2 + M. \tag{2.87}$$

Methanol (CH_3OH): 5 steps [19]

$$CH_3OH + M \rightarrow CH_2O + H_2 + M, \tag{2.88}$$

$$CH_2O + M \rightarrow CO + 2H + M, \tag{2.89}$$

$$CO + H_2O \rightarrow CO_2 + H_2, \tag{2.90}$$

$$3H_2 + O_2 \rightarrow 2H + 2H_2O, \tag{2.91}$$

$$H + H + M \rightarrow H_2 + M. \tag{2.92}$$

Methanol (CH_3OH): 6 steps [20]

$$CH_3OH + M \rightarrow CH_2O + H_2 + M, \tag{2.93}$$

$$CH_2O + M \rightarrow HCO + H + M, \tag{2.94}$$

$$HCO + H \rightarrow H_2 + CO, \tag{2.95}$$

$$CO + H_2O \rightarrow CO_2 + H_2, \tag{2.96}$$

$$3H_2 + O_2 \rightarrow 2H + 2H_2O, \tag{2.97}$$

$$H + H + M \rightarrow H_2 + M. \tag{2.98}$$

Ethanol (C_2H_5OH): 5 steps [21]

$$C_2H_5OH + M \rightarrow C_2H_4 + H_2O + M, \tag{2.99}$$

$$C_2H_4 + O_2 \rightarrow 2CO + 2H_2, \tag{2.100}$$

$$CO + H_2O \rightarrow CO_2 + H_2, \tag{2.101}$$

$$3H_2 + O_2 \rightarrow 2H + 2H_2O, \tag{2.102}$$

$$H + H + M \rightarrow H_2 + M. \tag{2.103}$$

Ethanol (C_2H_5OH): 7 steps [20]

$$2C_2H_5OH + H \rightarrow CH_2HCO + C_2H_4 + H_2O + 2H_2, \tag{2.104}$$

$$CH_2HCO + O_2 + H_2 \rightarrow CH_2O + CO + H_2O + H, \tag{2.105}$$

$$C_2H_4 + O_2 \rightarrow CH_2O + CO + H_2, \tag{2.106}$$

$$CH_2O + M \rightarrow CO + 2H + M, \tag{2.107}$$

$$CO + H_2O \rightarrow CO_2 + H_2, \tag{2.108}$$

$$3H_2 + O_2 \rightarrow 2H + 2H_2O, \tag{2.109}$$

$$H + H + M \rightarrow H_2 + M. \tag{2.110}$$

Propane (C_3H_8): 6 steps

$$2C_3H_8 + M \rightarrow 2C_3H_6 + 2H_2 + M, \tag{2.111}$$

$$C_3H_6 + H_2O \rightarrow C_2H_4 + CO + 2H_2, \tag{2.112}$$

$$C_2H_4 + O_2 + 2H \rightarrow 2CO + 3H_2, \tag{2.113}$$

$$CO + H_2O \rightarrow CO_2 + H_2, \tag{2.114}$$

$$3H_2 + O_2 \rightarrow 2H + 2H_2O, \tag{2.115}$$

$$H + H + M \rightarrow H_2 + M. \tag{2.116}$$

n-Heptane (nC_7H_{16}): 8 steps

$$nC_7H_{16} + O_2 + H_2 \rightarrow pC_4H_9 + C_2H_4 + CH_2O + H_2O + H, \tag{2.117}$$

$$pC_4H_9 + H \rightarrow 2C_2H_4 + H_2, \tag{2.118}$$

$$C_2H_4 + O_2 \rightarrow HCCO + H_2O + H, \tag{2.119}$$

$$HCCO + H_2O \rightarrow CH_2O + CO + H, \tag{2.120}$$

$$CH_2O + M \rightarrow CO + 2H + M, \tag{2.121}$$

$$CO + H_2O \rightarrow CO_2 + H_2, \tag{2.122}$$

$$3H_2 + O_2 \rightarrow 2H + 2H_2O, \tag{2.123}$$

$$H + H + M \rightarrow H_2 + M. \tag{2.124}$$

Iso-octane (iC_8H_{18}): 9 steps

$$iC_8H_{18} \rightarrow iC_4H_8 + pC_3H_4 + CH_3 + H_2 + H, \tag{2.125}$$

$$iC_4H_8 \rightarrow pC_3H_4 + CH_3 + H, \tag{2.126}$$

$$pC_3H_4 + CH_3 + O_2 \rightarrow C_2H_4 + HCCO + H_2O, \tag{2.127}$$

$$C_2H_4 + O_2 \rightarrow HCCO + H_2O + H, \tag{2.128}$$

$$HCCO + H_2O \rightarrow CH_2O + CO + H, \tag{2.129}$$

$$CH_2O + M \rightarrow CO + 2H + M, \tag{2.130}$$

$$CO + H_2O \rightarrow CO_2 + H_2, \tag{2.131}$$

$$3H_2 + O_2 \rightarrow 2H + 2H_2O, \tag{2.132}$$

$$H + H + M \rightarrow H_2 + M. \tag{2.133}$$

NO_x: 7 steps

$$CO + H \rightarrow CH + O, \tag{2.134}$$

$$N_2 + CH \rightarrow HCN + N, \tag{2.135}$$

$$N_2 + O + M \rightarrow N_2O + M, \tag{2.136}$$

$$HCN + 2O \rightarrow NO + CO + H, \tag{2.137}$$

$$N + O_2 \rightarrow NO + O, \tag{2.138}$$

$$N_2O + O \rightarrow 2NO, \tag{2.139}$$

$$4NO + 3O_2 \rightarrow 4NO_2 + 2O. \tag{2.140}$$

In Table 2.5 is shown a strong reduced mechanism for biodiesel, given here in terms of the methyl oleate, $C_{19}H_{36}O_2$. A strong reduction is desired to permit the solution of both problems, turbulence and chemical kinetics, with nonprohibitive computational times.

TABLE 2.5 Reduced Mechanism for Diffusion Flames of Biodiesel

Reaction	Reference
1. $H + H_2O = OH + H_2$	[22]
2. $H + O_2 + M = HO_2 + M$	[22]
3. $H_2 + OH = H_2O + H$	[22]
4. $O + H_2O = 2OH$	[22]
5. $O + OH = H + O_2$	[22]
6. $O_2 + H = OH + O$	[22]
7. $2OH = H_2O + O$	[22]
8. $HO_2 + OH = H_2O + O_2$	[22]
9. $2HO_2 = H_2O_2 + O_2$	[22]
10. $H_2O_2 + M = 2OH + M$	[22]
11. $CO + OH = CO_2 + H$	[22]
12. $H + CO_2 = OH + CO$	[22]
13. $CHO + M = CO + H + M$	[22]
14. $CHO + H = CO + H_2$	[22]
15. $CHO + OH = CO + H_2O$	[22]
16. $CH_2 + O_2 = CO + OH + H$	[22]
17. $CH_2O + OH = CHO + H_2O$	[22]
18. $CH_3 + O = CH_2O + H$	[22]
19. $CH_3 + O_2 = CH_2O + OH$	[22]
20. $CH_3OCO = CH_3 + CO_2$	[15]
21. $CH_4 + O_2 = CH_3 + HO_2$	[23]
22. $C_2H_2 + O = CH_2 + CO$	[22]
23. $C_2H_3 = C_2H_2 + H$	[22]
24. $C_2H_3 + O_2 = C_2H_2 + HO_2$	[22]
25. $C_2H_3 + O_2 = CH_2O + CHO$	[22]
26. $C_2H_4 + OH = C_2H_3 + H_2O$	[22]
27. $C_2H_3CO = C_2H_3 + CO$	[23]
28. $C_2H_5 + O_2 = C_2H_4 + HO_2$	[23]

Continued

TABLE 2.5 Reduced Mechanism for Diffusion Flames of Biodiesel—Cont'd

29. $C_3H_4 + HO_2 = C_2H_4 + CO + OH$	[15]
30. $C_3H_5 + O_2 = C_2H_2 + CH_2O + OH$	[15]
31. $C_3H_6 + CH_3 = C_3H_5 + CH_4$	[15]
32. $C_4H_6O_2 + HO_2 = C_2H_3CO + CH_2O + H_2O_2$	[23]
33. $C_5H_9O_2 = CH_3OCO + C_3H_6$	[15]
34. $C_5H_{10}O_2 + O_2 = HO_2 + C_5H_9O_2$	[23]
35. $C_5H_{10}O_2 + C_2H_3 = C_2H_4 + C_5H_9O_2$	[23]
36. $C_7H_{15} = C_2H_5 + C_2H_4 + C_3H_6$	[23]
37. $nC_7H_{16} + HO_2 = C_7H_{15} + H_2O_2$	[23]
38. $C_{11}H_{22}O_2 = C_4H_6O_2 + nC_7H_{16}$	[24]
39. $C_{19}H_{36}O_2 + O_2 = C_{11}H_{22}O_2 + C_5H_{10}O_2 + C_3H_4$	[24]

In the next chapter, the governing equations of flow are presented, which upon introduction/coupling with the equations of chemical kinetics produce the equations for reactive flows.

REFERENCES

[1] Vallance C. Reaction kinetics. Vallance group, Chemistry Research Laboratory, University of Oxford; 2012. Consulted in February 22, 2012, http://vallance.chem.ox.ac.uk/CVteaching.html.

[2] Fogler HS. Elements of chemical reaction engineering. Englewood Cliffs, NJ: Prentice Hall; 2005.

[3] Warnatz J, Maas U, Dibble RW. Combustion: physical and chemical fundamentals, modeling and simulation, experiments, pollutant formation. Berlin/Heidelberg: Springer-Verlag; 2006.

[4] Peters N. Fifteen lectures on laminar and turbulent combustion. Aachen, Germany: Ercoftac Summer School; 1992. Consulted in July 15, http://www.itv.rwth-aachen.de/fileadmin/LehreSeminar/Combustion/SummerSchool.pdf.

[5] Peters N. Turbulent combustion. UK: Cambridge University Press; 2006.

[6] Griffiths JF. Reduced kinetic models and their application to practical combustion systems. Prog Energy Combust Sci 1995;21:25-107.

[7] Curran HJ, Gaffuri P, Pitz WJ, Westbrook CK. A comprehensive modeling study of iso-octane oxidation. Combust Flame 2002;129:253-80.

[8] Atkins PPW, de Paula J. Atkin's physical chemistry. UK: Oxford University Press; 2006.

[9] Ebbing DD. General chemistry. Boston, MA: Houghton Mifflin; 1990.

[10] Lu T, Law CK. Linear time reduction of large kinetic mechanism with directed relation graph: n-heptane and iso-octane. Combust Flame 2006;144:24-36.

[11] Peters N, Rogg B. Reduced kinetic mechanisms for applications in combustion systems. In: Lecture notes in physics. Berlin/Heidelberg: Springer-Verlag; 1993.
[12] Westbrook CK, Mizobuchi Y, Poinsot TJ, Smith PJ, Warnatz J. Computational combustion. Proc Combust Inst 2005;30:125-57.
[13] Williams FA. Progress in knowledge of flamelet structure and extinction. Prog Energy Combust Sci 2000;26:657-82.
[14] Peters N. Systematic reduction of flame kinetics: principles and details. In: Dynamics of reactive systems. Part I: Flames. Progress in astronautics and aeronautics. Monmouth Junction: American Institute of Astronautics and Aeronautics; 1988.
[15] Mehl M, Curran HJ, Pitz WJ, Westbrook CK. ic8_ver3_mech.txt; 2009. Consulted in July 6, 2011, https://www-pls.llnl.gov/?url=science_and_technology-chemistry-combustionindexcombustion-prf.
[16] Glassmaker NJ. Intrinsic low-dimensional manifold method for rational simplification of chemical kinetics; 1999. p. 1-37. Consulted in July 13, 2010, http://www.nd.edu/~powers/nick.glassmaker.pdfindexpdf.
[17] Blanquart G, Pepiot-Desjardins P, Pitsch H. Chemical mechanism for high temperature combustion of engine relevant fuels with emphasis on soot precursors. Combust Flame 2009;156:588-607.
[18] Mauss F, Peters N, Rogg B, Williams FA. Reduced kinetic mechanism for premixed hydrogen flames. In: Reduced kinetic mechanisms for applications in combustion systems, Lecture notes in physics. Berlin/Heidelberg: Springer-Verlag; 1993.
[19] Müller CM, Seshadri K, Chen JY. Reduced kinetic mechanism for counterflow methanol diffusion flames. In: Reduced kinetic mechanisms for applications in combustion systems, Lecture notes in physics. Berlin/Heidelberg: Springer-Verlag; 1993.
[20] De Bortoli AL, Andreis GSL. Asymptotic analysis for coupled hydrogen, carbon monoxide, methanol and ethanol reduced kinetic mechanism. Lat Am Appl Res 2012;42:299-304.
[21] De Bortoli AL, Vaz FA, Lorenzzetti GS, Martins I. Systematic reduction of combustion reaction mechanism of common hydrocarbons and oxygenated fuels. Am Inst Phys Conf Proc ICNAAM 2010;1281:558-61.
[22] Seiser H, Pitsch H, Seshadri K, Pitz WJ, Curran HJ. Extinction and autoignition of n-heptane in counterflow configuration. Proc Combust Inst 2000;28:2029-37.
[23] Um S, Park SW. Numerical study on combustion and emission characteristics of homogeneous charge compression ignition engines fueled with biodiesel. Energy Fuels 2010;24:916-27. doi:10.1021/ef901092h.
[24] Golovitchev VI, Yang J. The construction of the combustion models for RME bio-diesel fuel for ice application; 2008. p. 1-15. Consulted in August 15, 2012, http://publications.lib.chalmers.se/records/fulltext/local_88943.pdf.

Chapter 3

Equations of Fluid Dynamics

The basic equations of fluid dynamics include the conservation of mass, momentum, and energy. The conservation of momentum is expressed by the Navier-Stokes equations, which determine the velocity field in a flow.

3.1 GOVERNING EQUATIONS FOR NONREACTIVE FLOWS

The Law of Conservation of a given property P, where CV corresponds to the control volume, is given by the following balance:

$$\left\{\begin{array}{l}\text{Rate of } P \text{ that}\\ \text{enters the CV}\end{array}\right\} - \left\{\begin{array}{l}\text{Rate of } P \text{ that}\\ \text{goes out}\\ \text{of the CV}\end{array}\right\} + \left\{\begin{array}{l}\text{Rate of } P\\ \text{generated}\\ \text{in the CV}\end{array}\right\} = \left\{\begin{array}{l}\text{Variation of}\\ P \text{ in the CV}\end{array}\right\}.$$

There is variation of P in the transient regime, that is, $\partial P/\partial t \neq 0$, while in the steady state there is no variation, that is, $\partial P/\partial t = 0$. In the following section, the equations of continuity, momentum, and energy are obtained.

3.1.1 Equation of Continuity

The principle of continuity is the conservation of overall mass,

$$\frac{\mathrm{D}}{\mathrm{D}t}\int_V \rho\,\mathrm{d}V = 0. \tag{3.1}$$

With the application of the Reynolds Transport Theorem, $\frac{\mathrm{D}p}{\mathrm{D}t} = \int_V \frac{\partial p}{\partial t}\,\mathrm{d}V + \int_S p(\vec{v}\cdot\vec{n})\,\mathrm{d}S$ and the Divergence Theorem, $\int_V \vec{\nabla}\cdot X\,\mathrm{d}V = \int_S X\cdot\vec{n}\,\mathrm{d}S$ (Appendix A), it follows that

$$\int_V \frac{\partial\rho}{\partial t}\,\mathrm{d}V + \int_S \rho v_j n_j\,\mathrm{d}S = 0 \Rightarrow \int_V \frac{\partial\rho}{\partial t}\,\mathrm{d}V + \int_V \frac{\partial(\rho v_j)}{\partial x_j}\,\mathrm{d}V = 0. \tag{3.2}$$

Because the control volume is time-invariant,

$$\frac{\partial\rho}{\partial t} + \frac{\partial(\rho v_j)}{\partial x_j} = 0 \Rightarrow \frac{\partial\rho}{\partial t} + \rho\frac{\partial v_j}{\partial x_j} + v_j\frac{\partial\rho}{\partial x_j} = 0, \tag{3.3}$$

Modeling and Simulation of Reactive Flows. http://dx.doi.org/10.1016/B978-0-12-802974-9.00003-9

and using the definition of material derivative, $\frac{D\vec{v}}{Dt} = \frac{\partial \vec{v}}{\partial t} + \vec{v} \cdot \vec{\nabla}\vec{v}$ (Appendix A), results in

$$\frac{D\rho}{Dt} + \rho\frac{\partial v_j}{\partial x_j} = 0. \tag{3.4}$$

For incompressible flows, ρ is constant along the flow, and $D\rho/Dt = 0$.

Equation (3.4) can also be obtained from the application of the Law of Mass Conservation on a differential element, in Cartesian coordinates for convenience, as shown in Figure 3.1.

After applying the overall mass balance to the control volume,

$$\left\{\begin{array}{l}\text{Mass that}\\ \text{enters the CV}\end{array}\right\} - \left\{\begin{array}{l}\text{Mass that goes}\\ \text{out of the CV}\end{array}\right\} = \left\{\begin{array}{l}\text{Mass accumulated}\\ \text{in the CV}\end{array}\right\}$$

the following expression is obtained

$$\begin{aligned}&\left[\rho v_x \,dy\,dz|_x + \rho v_y \,dx\,dz\big|_y + \rho v_z \,dx\,dy|_z\right]\\ &\quad - \left[\rho v_x \,dy\,dz|_{x+dx} + \rho v_y \,dx\,dz\big|_{y+dy} + \rho v_z \,dx\,dy|_{z+dz}\right] = \frac{\partial}{\partial t}(\rho \,dx\,dy\,dz).\end{aligned} \tag{3.5}$$

Rearranging terms results in

$$\begin{aligned}&\left[\rho v_x|_{x+dx} - \rho v_x|_x\right] dy\,dz + \left[\rho v_y\big|_{y+dy} - \rho v_y\big|_y\right] dx\,dz\\ &\quad + \left[\rho v_z|_{z+dz} - \rho v_z|_z\right] dx\,dy + \frac{\partial \rho}{\partial t} dx\,dy\,dz = 0.\end{aligned} \tag{3.6}$$

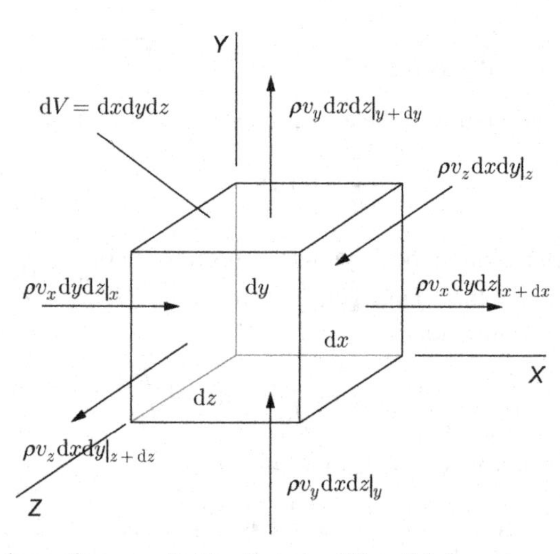

FIGURE 3.1 Balance of mass applied to a Cartesian differential element.

Dividing Eq. (3.6) by the volume variation $\mathrm{d}V = \mathrm{d}x\,\mathrm{d}y\,\mathrm{d}z$ and taking the limit when $\mathrm{d}V \to 0$ results the continuity equation in Cartesian coordinates (for cylindrical and spherical coordinates, see Appendix B).

$$\frac{\partial \rho}{\partial t} + \frac{\partial}{\partial x}(\rho v_x) + \frac{\partial}{\partial y}\left(\rho v_y\right) + \frac{\partial}{\partial z}(\rho v_z) = 0, \tag{3.7}$$

which is equivalent to Eq. (3.4). Expressing this equation in vector notation, results in

$$\frac{\partial \rho}{\partial t} + \vec{\nabla} \cdot (\rho \vec{v}) = 0. \tag{3.8}$$

The equation of continuity describes the density changes for a fixed point due to variations of the velocity vector $\vec{v}$.

3.1.2 Equation of Momentum

In accordance with Newton's Second Law, the rate of temporal variation of momentum is equal to the applied forces. Thus,

$$\frac{\mathrm{D}}{\mathrm{D}t}\int_V \rho v_i\,\mathrm{d}V = \int_S f_i\,\mathrm{d}S + \int_V \rho g_i\,\mathrm{d}V, \tag{3.9}$$

where f_i are surface forces and g_i is typically associated with the gravitational force (proportional to mass). Surface forces can be expressed in terms of the stress tensor as $f_i = \sigma_{ij} n_j$, where n_j is the normal vector. With the Reynolds Transport Theorem and the Divergence Theorem, it follows that:

$$\begin{aligned}
&\int_V \frac{\partial(\rho v_i)}{\partial t}\,\mathrm{d}V + \int_S \rho v_i v_j n_j\,\mathrm{d}S = \int_S \sigma_{ij} n_j\,\mathrm{d}S + \int_V \rho g_i\,\mathrm{d}V \\
&\Rightarrow \int_V \frac{\partial(\rho v_i)}{\partial t}\,\mathrm{d}V + \int_V \frac{\partial(\rho v_i v_j)}{\partial x_j}\,\mathrm{d}V = \int_V \frac{\partial \sigma_{ij}}{\partial x_j}\,\mathrm{d}V + \int_V \rho g_i\,\mathrm{d}V.
\end{aligned} \tag{3.10}$$

Because the control volume is invariant in time, the results become:

$$\begin{aligned}
&\frac{\partial(\rho v_i)}{\partial t} + \frac{\partial(\rho v_i v_j)}{\partial x_j} = \frac{\partial \sigma_{ij}}{\partial x_j} + \rho g_i \\
&\Rightarrow \rho\frac{\partial v_i}{\partial t} + v_i\frac{\partial \rho}{\partial t} + \rho v_i\frac{\partial v_j}{\partial x_j} + \rho v_j\frac{\partial v_i}{\partial x_j} + v_i v_j\frac{\partial \rho}{\partial x_j} = \frac{\partial \sigma_{ij}}{\partial x_j} + \rho g_i \\
&\Rightarrow v_i\left[\frac{\partial \rho}{\partial t} + v_j\frac{\partial \rho}{\partial x_j} + \rho\frac{\partial v_j}{\partial x_j}\right] + \rho\left[\frac{\partial v_i}{\partial t} + v_j\frac{\partial v_i}{\partial x_j}\right] = \frac{\partial \sigma_{ij}}{\partial x_j} + \rho g_i.
\end{aligned} \tag{3.11}$$

From the continuity equation, Eq. (3.4), and the definition of material derivative becomes the momentum equation,

$$\rho\frac{\mathrm{D}v_i}{\mathrm{D}t} = \frac{\partial \sigma_{ij}}{\partial x_j} + \rho g_i. \tag{3.12}$$

In fluid mechanics, it is common to separate the stress tensor as a sum of pressure and viscous stress, that is, $\sigma_{ij} = -p\delta_{ij} + \tau_{ij}$, where δ_{ij} is the Kronecker delta function. Thus, for the components x, y, and z, the Navier equations are written as:

$$\rho\left(\frac{\partial v_x}{\partial t} + v_x\frac{\partial v_x}{\partial x} + v_y\frac{\partial v_x}{\partial y} + v_z\frac{\partial v_x}{\partial z}\right) = -\frac{\partial p}{\partial x} + \frac{\partial \tau_{xx}}{\partial x} + \frac{\partial \tau_{yx}}{\partial y} + \frac{\partial \tau_{zx}}{\partial z} + \rho g_x, \tag{3.13}$$

$$\rho\left(\frac{\partial v_y}{\partial t} + v_x\frac{\partial v_y}{\partial x} + v_y\frac{\partial v_y}{\partial y} + v_z\frac{\partial v_y}{\partial z}\right) = -\frac{\partial p}{\partial y} + \frac{\partial \tau_{xy}}{\partial x} + \frac{\partial \tau_{yy}}{\partial y} + \frac{\partial \tau_{zy}}{\partial z} + \rho g_y, \tag{3.14}$$

$$\rho\left(\frac{\partial v_z}{\partial t} + v_x\frac{\partial v_z}{\partial x} + v_y\frac{\partial v_z}{\partial y} + v_z\frac{\partial v_z}{\partial z}\right) = -\frac{\partial p}{\partial z} + \frac{\partial \tau_{xz}}{\partial x} + \frac{\partial \tau_{yz}}{\partial y} + \frac{\partial \tau_{zz}}{\partial z} + \rho g_z. \tag{3.15}$$

The equations for Newtonian fluids are obtained by replacing the Stokes relations for the stresses, getting the Navier-Stokes equations. The Stokes relations, which are expressions for the shear and normal stresses as a function of velocity gradients, are given by

$$\tau_{xy} = \tau_{yx} = \mu\left(\frac{\partial v_x}{\partial y} + \frac{\partial v_y}{\partial x}\right), \tag{3.16}$$

$$\tau_{yz} = \tau_{zy} = \mu\left(\frac{\partial v_y}{\partial z} + \frac{\partial v_z}{\partial y}\right), \tag{3.17}$$

$$\tau_{xz} = \tau_{zx} = \mu\left(\frac{\partial v_x}{\partial z} + \frac{\partial v_z}{\partial x}\right), \tag{3.18}$$

$$\tau_{xx} = \mu\left[2\frac{\partial v_x}{\partial x} - \frac{2}{3}\left(\frac{\partial v_x}{\partial x} + \frac{\partial v_y}{\partial y} + \frac{\partial v_z}{\partial z}\right)\right], \tag{3.19}$$

$$\tau_{yy} = \mu\left[2\frac{\partial v_y}{\partial y} - \frac{2}{3}\left(\frac{\partial v_x}{\partial x} + \frac{\partial v_y}{\partial y} + \frac{\partial v_z}{\partial z}\right)\right], \tag{3.20}$$

$$\tau_{zz} = \mu\left[2\frac{\partial v_z}{\partial z} - \frac{2}{3}\left(\frac{\partial v_x}{\partial x} + \frac{\partial v_y}{\partial y} + \frac{\partial v_z}{\partial z}\right)\right], \tag{3.21}$$

where μ is the dynamic viscosity. The set of momentum equations for Newtonian fluids is written as

$$\rho\frac{\mathrm{D}v_i}{\mathrm{D}t} = -\frac{\partial p}{\partial x_i} + \frac{\partial \tau_{ij}}{\partial x_j} + \rho g_i, \tag{3.22}$$

where

$$\tau_{ij} = \mu\left(\frac{\partial v_i}{\partial x_j} + \frac{\partial v_j}{\partial x_i}\right) - \frac{2}{3}\mu\left(\frac{\partial v_k}{\partial x_k}\right)\delta_{ij}. \tag{3.23}$$

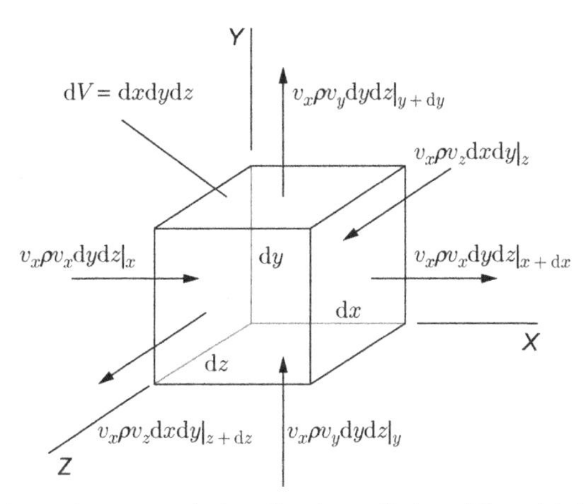

FIGURE 3.2 Balance of momentum in the x direction applied to a differential element in Cartesian coordinates.

For an incompressible Newtonian fluid, the following equation is used:

$$\rho \frac{\mathrm{D}v_i}{\mathrm{D}t} = -\frac{\partial p}{\partial x_i} + \mu \frac{\partial^2 v_i}{\partial x_j^2} + \rho g_i. \tag{3.24}$$

Equation (3.22) can also be obtained from the application of the Law of Conservation of momentum on a differential element, for the Cartesian direction x for convenience, as shown in Figure 3.2.

After applying the balance of conservation of momentum (CM) to this control volume,

$$\left\{\begin{array}{l}\text{Rate of CM that}\\ \text{enters the CV}\end{array}\right\} - \left\{\begin{array}{l}\text{Rate of CM that}\\ \text{goes out the CV}\end{array}\right\}$$
$$+ \left\{\begin{array}{l}\text{Rate of CM generated}\\ \text{inside the CV}\end{array}\right\} = \left\{\begin{array}{l}\text{Variation of CM}\\ \text{in the CV}\end{array}\right\}$$

the following expression results:

$$\begin{aligned} &-\Big\{\left[v_x\rho v_x|_{x+\mathrm{d}x} - v_x\rho v_x|_x\right] \mathrm{d}y\,\mathrm{d}z + \left[v_x\rho v_y\big|_{y+\mathrm{d}y} - v_x\rho v_y\big|_y\right] \mathrm{d}x\,\mathrm{d}z \\ &+ \left[v_x\rho v_z|_{z+\mathrm{d}z} - v_x\rho v_z|_z\right] \mathrm{d}x\,\mathrm{d}y\Big\} + \sum F_x = \frac{\partial}{\partial t}(\rho v_x)\,\mathrm{d}x\,\mathrm{d}y\,\mathrm{d}z. \end{aligned} \tag{3.25}$$

In the balance of momentum, the forces acting on the system $\left(\sum F_x\right)$ can be considered as a generation of momentum. Dividing Eq. (3.25) by volume $\mathrm{d}V = \mathrm{d}x\,\mathrm{d}y\,\mathrm{d}z$, and taking the limit when $\mathrm{d}V \to 0$, results in the equation of conservation momentum toward the x direction:

$$\frac{\partial}{\partial t}(\rho v_x) + \frac{\partial}{\partial x}(v_x\rho v_x) + \frac{\partial}{\partial y}\left(v_x\rho v_y\right) + \frac{\partial}{\partial z}(v_x\rho v_z) = \frac{\sum F_x}{\mathrm{d}V}. \tag{3.26}$$

Expanding this equation, one obtains

$$\rho\left(\frac{\partial v_x}{\partial t} + v_x\frac{\partial v_x}{\partial x} + v_y\frac{\partial v_x}{\partial y} + v_z\frac{\partial v_x}{\partial z}\right) \\ + v_x\left(\frac{\partial \rho}{\partial t} + \frac{\partial}{\partial x}(\rho v_x) + \frac{\partial}{\partial y}(\rho v_y) + \frac{\partial}{\partial z}(\rho v_z)\right) = \frac{\sum F_x}{\mathrm{d}V}, \tag{3.27}$$

where the second term within the parenthesis corresponds to the continuity equation, which has null value. Thus, Eq. (3.27) can be expressed as

$$\rho\left(\frac{\partial v_x}{\partial t} + v_x\frac{\partial v_x}{\partial x} + v_y\frac{\partial v_x}{\partial y} + v_z\frac{\partial v_x}{\partial z}\right) = \frac{\sum F_x}{\mathrm{d}V}. \tag{3.28}$$

The sum refers to the forces acting on the fluid element. These forces can be classified into two types:

- Surface forces, which are generated by contact with other fluid elements.
- Field forces acting throughout the control volume.

Surface forces act externally on the control volume, giving rise to the viscous stress τ on its surface. The body forces act throughout the control volume.

Figure 3.3 shows the surface forces acting on the fluid in the direction x on a Cartesian differential element.

Thus, the term $\sum F_x$ is written in terms of surface and body forces as

$$\sum F_x = \left[\tau_{xx}|_{x+\mathrm{d}x} - \tau_{xx}|_x\right] \mathrm{d}y\,\mathrm{d}z + \left[\tau_{xy}\big|_{y+\mathrm{d}y} - \tau_{xy}\big|_y\right] \mathrm{d}x\,\mathrm{d}z \\ + \left[\tau_{xz}|_{z+\mathrm{d}z} - \tau_{xz}|_z\right] \mathrm{d}x\,\mathrm{d}y - \left[p|_{x+\mathrm{d}x} - p|_x\right] \mathrm{d}y\,\mathrm{d}z + \rho g_x\,(\mathrm{d}x\,\mathrm{d}y\,\mathrm{d}z). \tag{3.29}$$

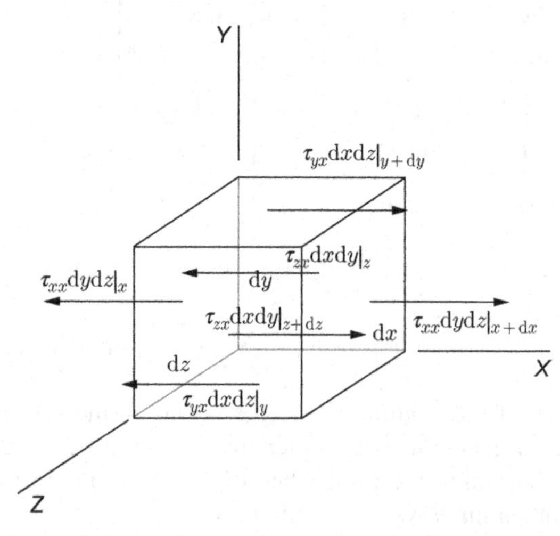

FIGURE 3.3 Balance of viscous forces acting on the x direction over a Cartesian differential element.

Dividing this equation by $\mathrm{d}V = \mathrm{d}x\,\mathrm{d}y\,\mathrm{d}z$ and taking the limit when $\mathrm{d}V \to 0$ results the equation

$$\frac{\sum F_x}{\mathrm{d}V} = \frac{\partial \tau_{xx}}{\partial x} + \frac{\partial \tau_{yx}}{\partial y} + \frac{\partial \tau_{zx}}{\partial z} - \frac{\partial p}{\partial x} + \rho g_x. \tag{3.30}$$

Combining Eq. (3.30) with Eq. (3.28) yields the equation that describes the conservation of momentum in the x direction in Cartesian coordinates (for cylindrical and spherical coordinates, see Appendix B).

$$\rho\left(\frac{\partial v_x}{\partial t} + v_x\frac{\partial v_x}{\partial x} + v_y\frac{\partial v_x}{\partial y} + v_z\frac{\partial v_x}{\partial z}\right) = -\frac{\partial p}{\partial x} + \frac{\partial \tau_{xx}}{\partial x} + \frac{\partial \tau_{yx}}{\partial y} + \frac{\partial \tau_{zx}}{\partial z} + \rho g_x. \tag{3.31}$$

Similarly, the equations for conservation of momentum in the directions y and z are obtained, which in vectorial notation have the form

$$\rho\left(\frac{\partial \vec{v}}{\partial t} + \vec{v}\cdot\vec{\nabla}\vec{v}\right) = -\vec{\nabla}p + \vec{\nabla}\cdot\vec{\vec{\tau}} + \rho\vec{g}. \tag{3.32}$$

The momentum equation tells us that a small element of volume that moves with the fluid can be accelerated by forces acting on it. In other words, the momentum equation corresponds to an expression of Newton's Second Law, whereby, *mass × acceleration = sum of forces*. Thus, the balance of momentum is equivalent to Newton's Second Law [1].

3.1.3 Energy Equation

According to the First Law of Thermodynamics, the variation of total energy of a system is equal to the rate of work done and heat entering through the boundaries,

$$\frac{\mathrm{D}E}{\mathrm{D}t} = \dot{W} + \dot{Q}, \tag{3.33}$$

where the total energy includes the internal energy, kinetic energy, and potential energy. For the deduction of the energy conservation equation, consider the parcel of the potential energy added to the field forces g_i. Thus,

$$\frac{\mathrm{D}}{\mathrm{D}t}\int_V \rho e\,\mathrm{d}V = \int_S f_i v_j\,\mathrm{d}S + \int_V \rho g_i v_j\,\mathrm{d}V + \int_S -\dot{q}_j n_j\,\mathrm{d}S + \int_V \dot{q}_\mathrm{v}\,\mathrm{d}V. \tag{3.34}$$

The first integral on the right side of Eq. (3.34) refers to forces on the surface of control volume, the second is the work of field forces, the third is the heat flow by conduction, and the last represents the volumetric sources of heat as the internal production of heat by Joule effect and chemical reactions.

With the definition of $f_i = \sigma_{ij}n_j$, the Reynolds Transport Theorem and the Divergence Theorem (Appendix A), it follows that:

$$
\begin{aligned}
&\int_V \frac{\partial}{\partial t}(\rho e)\,\mathrm{d}V + \int_S \rho e v_j n_j\,\mathrm{d}S \\
&= \int_S \sigma_{ij} v_j n_j\,\mathrm{d}S + \int_V \rho g_i v_j\,\mathrm{d}V + \int_S -\dot{q}_j n_j\,\mathrm{d}S + \int_V \dot{q}_\mathrm{v}\,\mathrm{d}V \\
&\Rightarrow \int_V \frac{\partial}{\partial t}(\rho e)\,\mathrm{d}V + \int_V \frac{\partial}{\partial x_j}\left(\rho e v_j\right)\mathrm{d}V \\
&= \int_V \frac{\partial(\sigma_{ij} v_j)}{\partial x_j}\,\mathrm{d}V + \int_V \rho g_i v_j\,\mathrm{d}V + \int_V \frac{\partial(-\dot{q}_j)}{\partial x_j}\,\mathrm{d}V + \int_V \dot{q}_\mathrm{v}\,\mathrm{d}V.
\end{aligned}
\tag{3.35}
$$

Because the control volume is time-invariant, the results are:

$$
\begin{aligned}
&\frac{\partial}{\partial t}(\rho e) + \frac{\partial}{\partial x_j}\left(\rho e v_j\right) = \frac{\partial(\sigma_{ij} v_j)}{\partial x_j} + \rho g_i v_j + \frac{\partial(-\dot{q}_j)}{\partial x_j} + \dot{q}_\mathrm{v} \\
&\Rightarrow \rho\frac{\partial e}{\partial t} + e\frac{\partial \rho}{\mathrm{d}t} + \rho v_j \frac{\partial e}{\partial x_j} + \rho e \frac{\partial v_j}{\partial x_j} + v_j e \frac{\partial \rho}{\partial x_j} \\
&= \sigma_{ij}\frac{\partial v_j}{\partial x_j} + v_j \frac{\partial \sigma_{ij}}{\partial x_j} + \rho g_i v_j + \frac{\partial(-\dot{q}_j)}{\partial x_j} + \dot{q}_\mathrm{v} \\
&\Rightarrow e\left(\frac{\partial \rho}{\mathrm{d}t} + v_j\frac{\partial \rho}{\partial x_j} + \rho\frac{\partial v_j}{\partial x_j}\right) + \rho\left(\frac{\partial e}{\partial t} + v_j\frac{\partial e}{\partial x_j}\right) \\
&= \sigma_{ij}\frac{\partial v_j}{\partial x_j} + v_j\left(\frac{\partial \sigma_{ij}}{\partial x_j} + \rho g_i\right) + \frac{\partial(-\dot{q}_j)}{\partial x_j} + \dot{q}_\mathrm{v}.
\end{aligned}
\tag{3.36}
$$

Using the continuity equation (3.4) and the definition of material derivative results in:

$$
\rho\frac{\mathrm{D}e}{\mathrm{D}t} = \sigma_{ij}\frac{\partial v_j}{\partial x_j} + \rho v_j \frac{\mathrm{D}v_i}{\mathrm{D}t} + \frac{\partial(-\dot{q}_j)}{\partial x_j} + \dot{q}_\mathrm{v}. \tag{3.37}
$$

Considering that the total energy of the fluid is the sum of the internal energy (e_int) and the kinetic energy ($e_\mathrm{cin} = v_j^2/2$), the following equation for the conservation of internal energy is obtained:

$$
\rho\frac{\mathrm{D}e_\mathrm{int}}{\mathrm{D}t} = \sigma_{ij}\frac{\partial v_j}{\partial x_j} + \frac{\partial(-\dot{q}_j)}{\partial x_j} + \dot{q}_\mathrm{v}. \tag{3.38}
$$

Introducing the Fourier's Law for heat conduction, $\dot{q}_j = -\kappa\partial T/\partial x_j$, yields the equation of internal energy in the form

$$
\rho\frac{\mathrm{D}e_\mathrm{int}}{\mathrm{D}t} = \sigma_{ij}\frac{\partial v_j}{\partial x_j} + \frac{\partial}{\partial x_j}\left(\kappa\frac{\partial T}{\partial x_j}\right) + \dot{q}_\mathrm{v}, \tag{3.39}
$$

where κ is the thermal conductivity. For a Newtonian fluid, this equation becomes

$$
\rho\frac{\mathrm{D}e_\mathrm{int}}{\mathrm{D}t} = \frac{\partial}{\partial x_j}\left(\kappa\frac{\partial T}{\partial x_j}\right) + \dot{q}_\mathrm{v} - p\left(\frac{\partial v_j}{\partial x_j}\right) + \Phi_{ij}, \tag{3.40}
$$

where Φ_{ij} is the viscous dissipation function given by

$$\Phi_{ij} = \frac{1}{2}\mu\left(\frac{\partial v_i}{\partial x_j} + \frac{\partial v_j}{\partial x_i}\right)^2 - \frac{2}{3}\mu\left(\frac{\partial v_j}{\partial x_j}\right)^2. \tag{3.41}$$

The viscous dissipation is often small, almost negligible, but it is important in flows of high speed and when the fluid viscosity is high.

Equation (3.40) can be written as

$$\begin{aligned}
&\rho\left[\frac{\partial e_{\text{int}}}{\partial t} + v_x\frac{\partial e_{\text{int}}}{\partial x} + v_y\frac{\partial e_{\text{int}}}{\partial y} + v_z\frac{\partial e_{\text{int}}}{\partial z}\right] \\
&= \frac{\partial}{\partial x}\left(\kappa\frac{\partial T}{\partial x}\right) + \frac{\partial}{\partial y}\left(\kappa\frac{\partial T}{\partial y}\right) + \frac{\partial}{\partial z}\left(\kappa\frac{\partial T}{\partial z}\right) + \dot{q}_{\text{v}} - p\left(\frac{\partial v_x}{\partial x} + \frac{\partial v_y}{\partial y} + \frac{\partial v_z}{\partial z}\right) \\
&\quad + \left\{2\mu\left[\left(\frac{\partial v_x}{\partial x}\right)^2 + \left(\frac{\partial v_y}{\partial y}\right)^2 + \left(\frac{\partial v_z}{\partial z}\right)^2\right]\right. \\
&\quad + \mu\left[\left(\frac{\partial v_x}{\partial y} + \frac{\partial v_y}{\partial x}\right)^2 + \left(\frac{\partial v_x}{\partial z} + \frac{\partial v_z}{\partial x}\right)^2 + \left(\frac{\partial v_y}{\partial z} + \frac{\partial v_z}{\partial y}\right)^2\right] \\
&\quad \left. -\frac{2}{3}\mu\left[\left(\frac{\partial v_x}{\partial x}\right)^2 + \left(\frac{\partial v_y}{\partial y}\right)^2 + \left(\frac{\partial v_z}{\partial z}\right)^2\right]\right\},
\end{aligned} \tag{3.42}$$

where the first term inside the braces is associated with the linear deformation of the fluid, the second to the angular deformation of the fluid, and the third to the dissipation due to the volumetric expansion of the fluid.

Equation (3.40) can also be obtained from the application of the Law of Conservation of energy on a differential element, Cartesian for convenience, as shown in Figure 3.4.

Applying the balance of energy conservation (E) to the control volume,

$$\left\{\begin{array}{l}\text{Rate of } E \text{ that} \\ \text{enters the CV}\end{array}\right\} - \left\{\begin{array}{l}\text{Rate of } E \text{ that} \\ \text{goes out of the CV}\end{array}\right\} + \left\{\begin{array}{l}\text{Rate of } E \text{ generated} \\ \text{inside the CV}\end{array}\right\} - \left\{\begin{array}{l}\text{Rate of} \\ \text{work done}\end{array}\right\} = \left\{\begin{array}{l}\text{Variation of } E \\ \text{inside the CV}\end{array}\right\},$$

the following expression is obtained

$$\begin{aligned}
&\left[(e\rho v_x + \dot{q}_x)|_{x+\text{d}x} - (e\rho v_x + \dot{q}_x)|_x\right]\text{d}y\,\text{d}z \\
&\quad + \left[(e\rho v_y + \dot{q}_y)\big|_{y+\text{d}y} - (e\rho v_y + \dot{q}_y)\big|_y\right]\text{d}x\,\text{d}z \\
&\quad + \left[(e\rho v_z + \dot{q}_z)|_{z+\text{d}z} - (e\rho v_z + \dot{q}_z)|_z\right]\text{d}x\,\text{d}y - \dot{q}_{\text{v}}\,\text{d}V + \frac{\partial}{\partial t}(e\rho)\,\text{d}V + \dot{W} = 0,
\end{aligned} \tag{3.43}$$

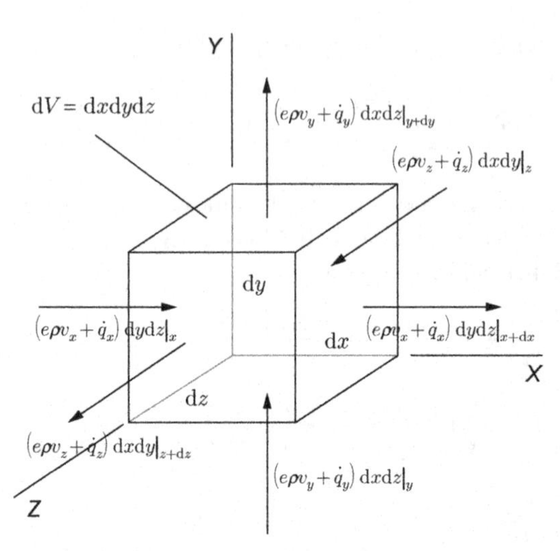

FIGURE 3.4 Balance of energy applied to a Cartesian differential element.

where e is the total energy per unit mass of the fluid element, $\vec{\dot{q}}$ the diffusive flux of heat through the control volume, $\dot{q}_{\text{v}}$ the rate of volumetric energy generation in the control volume, and $\dot{W}$ the work done by the fluid.

Dividing Eq. (3.43) by the volume $\mathrm{d}V = \mathrm{d}x\,\mathrm{d}y\,\mathrm{d}z$, and taking the limit when $\mathrm{d}V \to 0$, results the equation of conservation of energy

$$\frac{\partial}{\partial x}(e\rho v_x + \dot{q}_x) + \frac{\partial}{\partial y}\left(e\rho v_y + \dot{q}_y\right) + \frac{\partial}{\partial z}(e\rho v_z + \dot{q}_z) - \dot{q}_{\text{v}} + \frac{\dot{W}}{\mathrm{d}V} + \frac{\partial}{\partial t}(e\rho) = 0. \tag{3.44}$$

Expanding Eq. (3.44), one obtains

$$e\left[\frac{\partial \rho}{\partial t} + \frac{\partial}{\partial x}(\rho v_x) + \frac{\partial}{\partial y}\left(\rho v_y\right) + \frac{\partial}{\partial z}(\rho v_z)\right] + (\rho v_x)\frac{\partial e}{\partial x} + \left(\rho v_y\right)\frac{\partial e}{\partial y} + (\rho v_z)\frac{\partial e}{\partial z}$$
$$+\left(\frac{\partial \dot{q}_x}{\partial x} + \frac{\partial \dot{q}_y}{\partial y} + \frac{\partial \dot{q}_z}{\partial z}\right) - \dot{q}_{\text{v}} + \frac{\dot{W}}{\mathrm{d}V} + \rho\frac{\partial e}{\partial t} = 0, \tag{3.45}$$

where the term into brackets corresponds to the continuity equation. Thus, Eq. (3.45) can be expressed as

$$\rho\frac{\partial e}{\partial t} + (\rho v_x)\frac{\partial e}{\partial x} + \left(\rho v_y\right)\frac{\partial e}{\partial y} + (\rho v_z)\frac{\partial e}{\partial z}$$
$$+\left(\frac{\partial \dot{q}_x}{\partial x} + \frac{\partial \dot{q}_y}{\partial y} + \frac{\partial \dot{q}_z}{\partial z}\right) - \dot{q}_{\text{v}} + \frac{\dot{W}}{\mathrm{d}V} = 0. \tag{3.46}$$

Expressing this equation in vector notation, yields the energy equation

$$\rho \frac{\mathrm{D}e}{\mathrm{D}t} + \vec{\nabla} \cdot \vec{\dot{q}} - \dot{q}_{\mathrm{v}} = -\frac{\dot{W}}{\mathrm{d}V}. \tag{3.47}$$

In this equation, $-\dot{W}$ is equal to the rate of work that the environment exerts on the system, which is equal to the scalar product between the force applied and the velocity that exerts this force. Performing the balance of all the stresses exerted on the control volume yields

$$\begin{aligned}
-\dot{W} = & \left[\tau_{xx} v_x|_{x+\mathrm{d}x} - \tau_{xx} v_x|_x\right] \mathrm{d}y\,\mathrm{d}z + \left[\tau_{xy} v_y\big|_{x+\mathrm{d}x} - \tau_{xy} v_y\big|_x\right] \mathrm{d}y\,\mathrm{d}z \\
& + \left[\tau_{xz} v_z|_{x+\mathrm{d}x} - \tau_{xz} v_z|_x\right] \mathrm{d}y\,\mathrm{d}z + \left[\tau_{yx} v_x\big|_{y+\mathrm{d}y} - \tau_{yx} v_x\big|_y\right] \mathrm{d}x\,\mathrm{d}z \\
& + \left[\tau_{yy} v_y\big|_{y+\mathrm{d}y} - \tau_{yy} v_y\big|_y\right] \mathrm{d}x\,\mathrm{d}z + \left[\tau_{yz} v_z\big|_{y+\mathrm{d}y} - \tau_{yz} v_z\big|_y\right] \mathrm{d}x\,\mathrm{d}z \\
& + \left[\tau_{zx} v_x|_{z+\mathrm{d}z} - \tau_{zx} v_x|_z\right] \mathrm{d}x\,\mathrm{d}y + \left[\tau_{zy} v_y\big|_{z+\mathrm{d}z} - \tau_{zy} v_y\big|_z\right] \mathrm{d}x\,\mathrm{d}y \\
& + \left[\tau_{zz} v_z|_{z+\mathrm{d}z} - \tau_{zz} v_z|_z\right] \mathrm{d}x\,\mathrm{d}y - \left[p v_x|_{x+\mathrm{d}x} - p v_x|_x\right] \mathrm{d}y\,\mathrm{d}z \\
& - \left[p v_y\big|_{y+\mathrm{d}y} - p v_y\big|_y\right] \mathrm{d}x\,\mathrm{d}z - \left[p v_z|_{z+\mathrm{d}z} - p v_z|_z\right] \mathrm{d}x\,\mathrm{d}y \\
& + \rho v_x g_x \left(\mathrm{d}x\,\mathrm{d}y\,\mathrm{d}z\right) + \rho v_y g_y \left(\mathrm{d}x\,\mathrm{d}y\,\mathrm{d}z\right) + \rho v_z g_z \left(\mathrm{d}x\,\mathrm{d}y\,\mathrm{d}z\right).
\end{aligned} \tag{3.48}$$

Dividing this equation by $\mathrm{d}V = \mathrm{d}x\,\mathrm{d}y\,\mathrm{d}z$ and taking the limit when $\mathrm{d}V \to 0$ yields the equation

$$\begin{aligned}
-\frac{\dot{W}}{\mathrm{d}V} = & \frac{\partial}{\partial x}\left(\tau_{xx} v_x\right) + \frac{\partial}{\partial x}\left(\tau_{xy} v_y\right) + \frac{\partial}{\partial x}\left(\tau_{xz} v_z\right) + \frac{\partial}{\partial y}\left(\tau_{yx} v_x\right) + \frac{\partial}{\partial y}\left(\tau_{yy} v_y\right) \\
& + \frac{\partial}{\partial y}\left(\tau_{yz} v_z\right) + \frac{\partial}{\partial z}\left(\tau_{zx} v_x\right) + \frac{\partial}{\partial z}\left(\tau_{zy} v_y\right) + \frac{\partial}{\partial z}\left(\tau_{zz} v_z\right) - \frac{\partial}{\partial x}\left(p v_x\right) \\
& - \frac{\partial}{\partial y}\left(p v_y\right) - \frac{\partial}{\partial z}\left(p v_z\right) + \rho v_x g_x + \rho v_y g_y + \rho v_z g_z.
\end{aligned} \tag{3.49}$$

Expanding the derivatives, one obtains

$$\begin{aligned}
-\frac{\dot{W}}{\mathrm{d}V} = & v_x \left(\frac{\partial \tau_{xx}}{\partial x} + \frac{\partial \tau_{yx}}{\partial y} + \frac{\partial \tau_{zx}}{\partial z} - \frac{\partial p}{\partial x} + \rho g_x\right) \\
& + v_y \left(\frac{\partial \tau_{xy}}{\partial x} + \frac{\partial \tau_{yy}}{\partial y} + \frac{\partial \tau_{zy}}{\partial z} - \frac{\partial p}{\partial y} + \rho g_y\right) \\
& + v_z \left(\frac{\partial \tau_{xz}}{\partial x} + \frac{\partial \tau_{yz}}{\partial y} + \frac{\partial \tau_{zz}}{\partial z} - \frac{\partial p}{\partial z} + \rho g_z\right) \\
& + \left(\tau_{xx}\frac{\partial v_x}{\partial x} + \tau_{yx}\frac{\partial v_x}{\partial y} + \tau_{zx}\frac{\partial v_x}{\partial z}\right) + \left(\tau_{xy}\frac{\partial v_y}{\partial x} + \tau_{yy}\frac{\partial v_y}{\partial y} + \tau_{zy}\frac{\partial v_y}{\partial z}\right) \\
& + \left(\tau_{xz}\frac{\partial v_z}{\partial x} + \tau_{yz}\frac{\partial v_z}{\partial y} + \tau_{zz}\frac{\partial v_z}{\partial z}\right) - p\left(\frac{\partial v_x}{\partial x} + \frac{\partial v_y}{\partial y} + \frac{\partial v_z}{\partial z}\right).
\end{aligned} \tag{3.50}$$

On the right side of Eq. (3.50), the term $\left(\frac{\partial \tau_{xxj}}{\partial x} + \frac{\partial \tau_{yxj}}{\partial y} + \frac{\partial \tau_{zxj}}{\partial z} - \frac{\partial p}{\partial x_j} + \rho g_j\right)$ is equal to $\rho \frac{Dv_j}{Dt}$ according to the momentum equation. Thus, the expression for the work done by the fluid per unit volume becomes

$$-\frac{\dot{W}}{\mathrm{d}V} = \rho \left(v_x \frac{\mathrm{D}v_x}{\mathrm{D}t} + v_y \frac{\mathrm{D}v_y}{\mathrm{D}t} + v_z \frac{\mathrm{D}v_z}{\mathrm{D}t} \right) + A, \tag{3.51}$$

where A is given by

$$\begin{aligned} A = &\left(\tau_{xx} \frac{\partial v_x}{\partial x} + \tau_{yx} \frac{\partial v_x}{\partial y} + \tau_{zx} \frac{\partial v_x}{\partial z} \right) + \left(\tau_{xy} \frac{\partial v_y}{\partial x} + \tau_{yy} \frac{\partial v_y}{\partial y} + \tau_{zy} \frac{\partial v_y}{\partial z} \right) \\ &+ \left(\tau_{xz} \frac{\partial v_z}{\partial x} + \tau_{yz} \frac{\partial v_z}{\partial y} + \tau_{zz} \frac{\partial v_z}{\partial z} \right) - p \left(\frac{\partial v_x}{\partial x} + \frac{\partial v_y}{\partial y} + \frac{\partial v_z}{\partial z} \right). \end{aligned} \tag{3.52}$$

Expressing Eq. (3.51) in terms of the kinetic energy of the fluid $e_{\text{cin}} = (v_x^2 + v_y^2 + v_z^2)/2$ gives the following equation for work per unit volume exerted by the fluid

$$-\frac{\dot{W}}{\mathrm{d}V} = \rho \frac{\mathrm{D}e_{\text{cin}}}{\mathrm{D}t} + A. \tag{3.53}$$

Considering that the total energy of the fluid is the sum of internal energy and the kinetic energy ($e = e_{\text{int}} + e_{\text{cin}}$) and replacing Eq. (3.53) in Eq. (3.47) yields the following equation for internal energy:

$$\rho \frac{\mathrm{D}e_{\text{int}}}{\mathrm{D}t} = -\vec{\nabla} \cdot \vec{q} - p \left(\vec{\nabla} \cdot \vec{v} \right) + \vec{\vec{\tau}} : \left(\vec{\nabla} \vec{v} \right) + \dot{q}_{\text{v}}, \tag{3.54}$$

which is equivalent to Eq. (3.40).

For Eq. (3.54), the following physical interpretation for each of the terms can be made:

1. The first term corresponds to the change of internal energy per unit volume.
2. The second term is the gain of internal energy per unit volume due to thermal diffusion.
3. The third term indicates the work associated with the expansion and compression of the fluid (a reversible work).
4. The fourth term expresses the work associated with the viscous dissipation due to friction between the fluid layers, which transforms mechanical energy into thermal energy (irreversible).
5. The fifth term corresponds to the volumetric energy generation.

The equation for the conservation of internal energy can be written for Newtonian fluids in terms of specific enthalpy ($h = e_{\text{int}} + p/\rho$) as

$$\rho \frac{\mathrm{D}h}{\mathrm{D}t} = \rho \frac{\mathrm{D}e_{\text{int}}}{\mathrm{D}t} + \frac{\mathrm{D}p}{\mathrm{D}t} - \frac{p}{\rho} \frac{\mathrm{D}\rho}{\mathrm{D}t}, \tag{3.55}$$

and using the equation of energy for a Newtonian fluid (3.40) and the equation of continuity (3.4) results in

$$\rho\frac{\mathrm{D}h}{\mathrm{D}t}=\frac{\mathrm{D}p}{\mathrm{D}t}+\frac{\partial}{\partial x_j}\left(\kappa\frac{\partial T}{\partial x_j}\right)+\dot{q}_\mathrm{v}+\Phi_{ij}, \tag{3.56}$$

which in vectorial notation becomes

$$\rho\frac{\mathrm{D}h}{\mathrm{D}t}=\frac{\mathrm{D}p}{\mathrm{D}t}-\vec{\nabla}\cdot\vec{q}+\dot{q}_\mathrm{v}+\vec{\vec{\tau}}:\left(\vec{\nabla}\vec{v}\right). \tag{3.57}$$

The equation of energy conservation can also be written in terms of temperature. For this, consider that specific enthalpy is a function of pressure and temperature,

$$\mathrm{d}h=\left.\frac{\partial h}{\partial T}\right|_\mathrm{P}\mathrm{d}T+\left.\frac{\partial h}{\partial p}\right|_\mathrm{T}\mathrm{d}p=c_\mathrm{p}\,\mathrm{d}T+\frac{V}{\rho}(1-\beta T)\,\mathrm{d}p, \tag{3.58}$$

where c_p is the specific heat at constant pressure, $\frac{V}{\rho}$ the specific volume, and β the coefficient of thermal expansion $[\beta=-(\partial\rho/\partial T|_\mathrm{p})/\rho]$. After applying the material derivative and multiplying the equation by ρ, it follows that

$$\rho\frac{\mathrm{D}h}{\mathrm{D}t}=\rho c_\mathrm{p}\frac{\mathrm{D}T}{\mathrm{D}t}+(1-\beta T)\frac{\mathrm{D}p}{\mathrm{D}t}. \tag{3.59}$$

Substituting Eq. (3.59) in Eq. (3.55) yields the equation of energy for a Newtonian fluid in terms of temperature:

$$\rho c_\mathrm{p}\frac{\mathrm{D}T}{\mathrm{D}t}=\frac{\partial}{\partial x_j}\left(\kappa\frac{\partial T}{\partial x_j}\right)+\beta T\frac{\mathrm{D}p}{\mathrm{D}t}+\dot{q}_\mathrm{v}+\Phi_{ij}, \tag{3.60}$$

which in vector notation takes the form (for cylindrical and spherical coordinates, see Appendix B)

$$\rho c_\mathrm{p}\frac{\mathrm{D}T}{\mathrm{D}t}=-\vec{\nabla}\cdot\vec{q}+\beta T\frac{\mathrm{D}p}{\mathrm{D}t}+\dot{q}_\mathrm{v}+\vec{\vec{\tau}}:\left(\vec{\nabla}\vec{v}\right). \tag{3.61}$$

Simplified forms of the temperature equation can be obtained. For ideal gases, the thermal expansion coefficient is equal to $1/T$, so the equation for the temperature becomes

$$\rho c_\mathrm{p}\frac{\mathrm{D}T}{\mathrm{D}t}=\frac{\partial}{\partial x_j}\left(\kappa\frac{\partial T}{\partial x_j}\right)+\frac{\mathrm{D}p}{\mathrm{D}t}+\dot{q}_\mathrm{v}+\Phi_{ij}. \tag{3.62}$$

For incompressible fluids, $\beta=0$ in Eq. (3.60), resulting

$$\rho c_\mathrm{p}\frac{\mathrm{D}T}{\mathrm{D}t}=\frac{\partial}{\partial x_j}\left(\kappa\frac{\partial T}{\partial x_j}\right)+\dot{q}_\mathrm{v}+\Phi_{ij}. \tag{3.63}$$

This equation, when the thermal conductivity is constant and the viscous dissipation function is negligible, becomes

$$\frac{\mathrm{D}T}{\mathrm{D}t}=\alpha\frac{\partial^2 T}{\partial x_j^2}+\frac{\dot{q}_\mathrm{v}}{\rho c_\mathrm{p}}, \tag{3.64}$$

where $\alpha = \kappa/(\rho c_\mathrm{p})$ is the thermal diffusivity of the fluid. Expanding the terms of this equation yields

$$\frac{\partial T}{\partial t} + v_x\frac{\partial T}{\partial x} + v_y\frac{\partial T}{\partial y} + v_z\frac{\partial T}{\partial z} = \alpha\left(\frac{\partial^2 T}{\partial x^2} + \frac{\partial^2 T}{\partial y^2} + \frac{\partial^2 T}{\partial z^2}\right) + \frac{\dot{q}_\mathrm{v}}{\rho c_\mathrm{p}}. \tag{3.65}$$

In summary, the following set of equations for Newtonian fluid (continuity, momentum, and energy) are as follows:

$$\frac{\partial \rho}{\partial t} + \vec{\nabla}\cdot(\rho\vec{v}) = 0, \tag{3.66}$$

$$\rho\left(\frac{\partial \vec{v}}{\partial t} + \vec{v}\cdot\vec{\nabla}\vec{v}\right) = -\vec{\nabla}p + \vec{\nabla}\cdot\vec{\vec{\tau}} + \rho\vec{g}, \tag{3.67}$$

$$\rho\frac{\mathrm{D}e_\mathrm{int}}{\mathrm{D}t} = -\vec{\nabla}\cdot\vec{q} - p\left(\vec{\nabla}\cdot\vec{v}\right) + \vec{\vec{\tau}} : \left(\vec{\nabla}\vec{v}\right) + \dot{q}_\mathrm{v}, \tag{3.68}$$

where the strain term for a Newtonian fluid is given by

$$\vec{\vec{\tau}} = \mu\left[\vec{\nabla}\vec{v} + \left(\vec{\nabla}\vec{v}\right)^T - \frac{2}{3}\left(\vec{\nabla}\cdot\vec{v}\right)\vec{\vec{I}}\right], \tag{3.69}$$

where μ is the dynamic viscosity, and where $\vec{\vec{I}}$ corresponds to the identity tensor.

3.2 GOVERNING EQUATIONS FOR REACTIVE FLOWS

For reactive flows, in addition to the equations given in the previous section, the equation of conservation of mass for each chemical species i (mass fraction) appears. The continuity equation written for a single component is given by [2, 3]

$$\frac{\partial \rho_i}{\partial t} + \frac{\partial}{\partial x_j}[\rho_i(v_j)_i] = \dot{w}_i, \tag{3.70}$$

where ρ_i is the density of species i and $(v_j)_i$ the speed of species i. The volumetric rate of generation of species i via chemical reaction is given by

$$\dot{w}_i = W_i\sum_{k=1}^{r}\nu_{ik}\dot{w}_k, \tag{3.71}$$

where W_i is the molecular weight of species i, ν_{ik} the stoichiometric coefficient of the species i in the reaction k, and $\dot{w}_k$ the rate of reaction k in a mechanism containing r chemical reactions, given by $\dot{w}_k = k_{fk}\prod_{i=1}^{n}\left(\frac{\rho Y_i}{W_i}\right)^{\nu_{fik}} - k_{bk}\prod_{i=1}^{n}\left(\frac{\rho Y_i}{W_i}\right)^{\nu_{bik}}$. There is one equation for each species, and the sum of all these equations is

$$\frac{\partial}{\partial t}\left(\sum_{i=1}^{n}\rho_i\right) + \frac{\partial}{\partial x_j}\left[\sum_{i=1}^{n}\rho_i(v_j)_i\right] = \sum_{i=1}^{n}\dot{w}_i, \tag{3.72}$$

where n is the total number of species. Because $\sum_{i=1}^{n} \rho_i = \rho$, $\sum_{i=1}^{n} \rho_i (v_j)_i = \rho v_j$, where v_j is the mass average velocity defined by $v_j = \sum_{i=1}^{n} \rho_i (v_j)_i / \rho$, and $\sum_{i=1}^{n} \dot{w}_i = 0$. The equation (3.72) results in the continuity equation (3.4). In these equations $\rho_i (v_j)_i$ is the mass flow of species i; it is a result of advective and diffusive contributions, $\rho_i (v_j)_i = \rho_i v_j + j_i$, where the mass flow of species i due to diffusion (Fick's Law of Diffusion) is given by

$$j_i = -D_{ik} \frac{\partial (\rho Y_k)}{\partial x_j} \tag{3.73}$$

and D_{ik} is the mass diffusivity of species i through k species. Thus, Eq. (3.70) becomes

$$\frac{\partial \rho_i}{\partial t} + \frac{\partial}{\partial x_j} (\rho_i v_j + j_i) = \dot{w}_i. \tag{3.74}$$

Introducing the mass fraction $Y_i = \rho_i / \rho$ yields

$$\frac{\partial (\rho Y_i)}{\partial t} + \frac{\partial}{\partial x_j} (\rho Y_i v_j) = -\frac{\partial j_i}{\partial x_j} + \dot{w}_i. \tag{3.75}$$

Expanding these derivatives and using the continuity equation, Eq. (3.4), yields the equation of conservation of mass fractions

$$\rho \frac{\mathrm{D} Y_i}{\mathrm{D} t} = -\frac{\partial j_i}{\partial x_j} + \dot{w}_i, \quad i = 1, 2, \ldots, n, \tag{3.76}$$

which in vector notation is given by (for cylindrical and spherical coordinates see Appendix B)

$$\rho \frac{\mathrm{D} Y_i}{\mathrm{D} t} = -\vec{\nabla} \cdot \vec{j}_i + \dot{w}_i, \quad i = 1, 2, \ldots, n. \tag{3.77}$$

Usually, for reactive flows the choice should be the energy equation in the form of enthalpy or temperature due to numerical stability reasons. According to Peters [4], the equation for the enthalpy can be written as

$$\rho \frac{\mathrm{D} h}{\mathrm{D} t} = \frac{\mathrm{D} p}{\mathrm{D} t} - \vec{\nabla} \cdot \vec{q} + q_{\mathrm{R}}, \tag{3.78}$$

where the heat flux $\vec{q}$ includes the effect of enthalpy transport for diffusive flows $\vec{j}_i$, or

$$\vec{q} = -\kappa \vec{\nabla} T + \sum_{i=1}^{n} (h_i \vec{j}_i), \tag{3.79}$$

and q_{R} is the heat transfer due to radiation.

The energy equation can be written in terms of temperature as [4]

$$\rho c_{\mathrm{p}} \frac{\mathrm{D} T}{\mathrm{D} t} = \frac{\partial p}{\partial t} + \vec{\nabla} \cdot (\kappa \vec{\nabla} T) - \sum_{i=1}^{n} (c_{\mathrm{p}_i} \vec{j}_i \cdot \vec{\nabla} T) - \sum_{i=1}^{n} (h_i \dot{w}_i) + q_{\mathrm{R}}. \tag{3.80}$$

Considering that c_{p_i} is equal for all species i, pressure is constant, and the heat transfer due to radiation is negligible, one obtains

$$\rho \frac{\mathrm{D}T}{\mathrm{D}t} = \vec{\nabla} \cdot \left(\frac{\kappa}{c_\mathrm{p}} \vec{\nabla} T \right) + \dot{w}_\mathrm{T}, \tag{3.81}$$

where the heat loss due to chemical reactions is given by

$$\dot{w}_\mathrm{T} = -\frac{1}{c_\mathrm{p}} \sum_{i=1}^{n} (h_i \dot{w}_i). \tag{3.82}$$

Thus, the set of equations for reactive flows for Newtonian fluids can be written as (continuity, momentum, energy, and mass fraction)

$$\frac{\partial \rho}{\partial t} + \vec{\nabla} \cdot (\rho \vec{v}) = 0, \tag{3.83}$$

$$\rho \left(\frac{\partial \vec{v}}{\partial t} + \vec{v} \cdot \vec{\nabla} \vec{v} \right) = -\vec{\nabla} p + \vec{\nabla} \cdot \vec{\vec{\tau}} + \rho \vec{g}, \tag{3.84}$$

$$\rho \frac{\mathrm{D}T}{\mathrm{D}t} = \vec{\nabla} \cdot \left(\frac{\kappa}{c_\mathrm{p}} \vec{\nabla} T \right) + \dot{w}_\mathrm{T}, \tag{3.85}$$

$$\rho \frac{\mathrm{D}Y_i}{\mathrm{D}t} = -\vec{\nabla} \cdot \vec{j}_i + \dot{w}_i, \quad i = 1, 2, \ldots, n \tag{3.86}$$

with $\vec{\vec{\tau}}$ given by Eq. (3.69), $\vec{j}_i$ using Eq. (3.73), $\dot{w}_i$ by Eq. (3.71), and $\dot{w}_\mathrm{T}$ by Eq. (3.82).

The thermal conductivity, κ, and the dynamic viscosity of the fluid, μ, can be modeled using the arithmetic and harmonic means of each component in accordance with the expressions [5]:

$$\kappa = \frac{1}{2} \sum_{k=1}^{n} x_k \kappa_k + \frac{1}{2} \left(\sum_{k=1}^{n} \frac{x_k}{\kappa_k} \right)^{-1}, \tag{3.87}$$

$$\mu = \frac{1}{2} \sum_{k=1}^{n} x_k \mu_k + \frac{1}{2} \left(\sum_{k=1}^{n} \frac{x_k}{\mu_k} \right)^{-1}. \tag{3.88}$$

Numerical simulations of compressible reactive flows require a fine mesh resolution and long integration times to capture details of the flow. Compressible flows with high Mach numbers have high gradients and discontinuities, requiring the application of numerical simulations using high-order methods [6]. However, in most applications involving reactive flow, such as in burners and reactors, the speeds are low compared to the speed of sound, so the flow is hydrodynamically incompressible. In this case, the compressibility effects are from variations in temperature and pressure. The compressibility caused by the temperature variation is significantly more important than the contribution due to pressure variation. This leads to numerical problems when applying the compressible form of the Navier-Stokes equations for low Mach number [7].

REFERENCES

[1] Bird RB, Stewart WE, Lightfoot EN. Transport phenomena. 2nd ed. New York: John Wiley & Sons; 2006.
[2] Poinsot T, Veynante D. Theoretical and numerical combustion. Philadelphia, PA: R. T. Edwards, Inc.; 2001.
[3] Law CK. Combustion physics. New York: Cambridge University Press; 2006.
[4] Peters N. Turbulent combustion. New York: Cambridge University Press; 2006.
[5] Jäger W, Rannacher R, Warnatz J. Reactive flows, diffusion and transport: from experiments via mathematical modeling to numerical simulation and optimization. Berlin/Heidelberg: Springer; 2006.
[6] Gottlieb D, Gottlieb S. Spectral methods for compressible reactive flows. C R Mécanique 2005;333:3-16.
[7] Braack M. An adaptive finite element method for reactive flow problems. PhD thesis. Heidelberg University; 1998.

Chapter 4

Mixing and Turbulent Flows

Flows involving mixing and reaction are common in nature and have numerous applications of technical interest, such as combustion and geochemistry. In this chapter, some important definitions are introduced.

4.1 MIXTURE FRACTION

The mixture fraction is a variable frequently used for mixtures of two or three fluids. This fraction measures the mixture of reagents and is primarily concerned with the movements of large scales of flow [1].

Consider a homogeneous system in the absence of diffusion and the following overall reaction for the complete combustion of a hydrocarbon C_mH_n

$$\nu'_F C_m H_n + \nu'_{O_2} O_2 \rightarrow \nu''_{CO_2} CO_2 + \nu''_{H_2O} H_2O, \tag{4.1}$$

where ν'_F is the stoichiometric coefficient of the fuel, and the stoichiometric coefficient of the oxidant is given by $\nu'_{O_2} = (m + n/4)/\nu'_F$. For this reaction, Eq. (4.1) results the following relation between the variation of the mass fractions of oxygen dY_{O_2} and fuel dY_F:

$$\frac{dY_{O_2}}{\nu'_{O_2} W_{O_2}} = \frac{dY_F}{\nu'_F W_F}, \tag{4.2}$$

where W_i is the molecular weight of species i. Equation (4.2) can be integrated resulting in the equation

$$\nu Y_F - Y_{O_2} = \nu Y_{F,u} - Y_{O_2,u}, \tag{4.3}$$

where $\nu = \nu'_{O_2} W_{O_2}/(\nu'_F W_F)$ is the stoichiometric mass ratio of oxygen fuel, and the subscript u denotes the initial conditions in the unburned mixture. The mass fractions Y_F and Y_{O_2} correspond to any state of combustion between the burned and unburned state. For equal diffusivities, Eq. (4.3) is also valid in spatially inhomogeneous systems, as in the case of diffusion flames.

Modeling and Simulation of Reactive Flows. http://dx.doi.org/10.1016/B978-0-12-802974-9.00004-0

For a stoichiometric mixture, fuel and oxygen are completely consumed. The ratio of the concentrations of oxygen and unburned fuel is equal to the ratio of the stoichiometric coefficients, or

$$\left.\frac{[X_{O_2}]_u}{[X_F]_u}\right|_{st} = \frac{\nu'_{O_2}}{\nu'_F}. \tag{4.4}$$

Because the concentration of a species X_i is given by

$$[X_i] = \frac{\rho Y_i}{W_i}, \tag{4.5}$$

where ρ is the density, and $\nu = \nu'_{O_2} W_{O_2}/(\nu'_F W_F)$, it follows that

$$\nu = \left.\frac{Y_{O_2,u}}{Y_{F,u}}\right|_{st}. \tag{4.6}$$

Consider a system with two feeders, where a fuel mass flow $\dot{m}_1$ is mixed with an oxidant stream of mass flow $\dot{m}_2$. The mixture fraction Z is the mass fraction of the fuel stream in the mixture, or

$$Z = \frac{\dot{m}_1}{\dot{m}_1 + \dot{m}_2}. \tag{4.7}$$

Both the fuel stream and the oxidant stream may contain inerts, such as nitrogen. The mass fraction of fuel in the unburned mixture is related to the mixture fraction through the relation

$$Y_{F,u} = Y_{F,1} Z, \tag{4.8}$$

where $Y_{F,1}$ denotes the mass fraction of fuel in the fuel stream. Because $(1 - Z)$ is the mass fraction of oxidant stream in the mixture, the mass fraction of oxygen in the unburned mixture is given by

$$Y_{O_2,u} = Y_{O_2,2}(1 - Z), \tag{4.9}$$

where $Y_{O_2,2}$ is the mass fraction of oxygen in the oxidant stream, and $Y_{O_2,2} = 0.232$ for air. Substituting Eqs. (4.8) and (4.9) in (4.3), and integrating between the unburned state and any other state of combustion, the mixture fraction can be written as a function of the mass fractions of fuel and oxygen,

$$Z = \frac{\nu Y_F - Y_{O_2} + Y_{O_2,2}}{\nu Y_{F,1} + Y_{O_2,2}}. \tag{4.10}$$

In the case of a stoichiometric mixture, Eq. (4.3) becomes

$$\nu Y_F - Y_{O_2} = 0 \tag{4.11}$$

and therefore the stoichiometric mixture fraction is given by

$$Z_{st} = \left(1 + \frac{\nu Y_{F,1}}{Y_{O_2,2}}\right)^{-1}. \tag{4.12}$$

The use of mixture fraction rather than mass fraction (or concentration) facilitates the resolution of the system of equations, and from the mixture fraction the mass fractions of the components of the overall reaction can be determined.

The equivalence ratio ϕ is also a widely employed variable, defined as the ratio of fuel to air in the unburned mixture, normalized by the fuel-air ratio in the stoichiometric mixture,

$$\phi = \frac{Y_{F,u}/Y_{O_2,u}}{(Y_{F,u}/Y_{O_2,u})_{st}} = \frac{\nu Y_{F,u}}{Y_{O_2,u}}. \tag{4.13}$$

Substituting Eqs. (4.8) and (4.9) in Eq. (4.13), and using Eq. (4.12), results a relation between the equivalence ratio and the mixture fraction, which is given by

$$\phi = \left(\frac{Z}{1-Z}\right)\left(\frac{1-Z_{st}}{Z_{st}}\right). \tag{4.14}$$

This expression suggests that the mixture fraction can be interpreted as a normalized fuel-air equivalence ratio [1].

The mixture fraction can also be defined in a more general way. First, an equation for the mass fraction of elements is defined. While the mass of the chemical species vary due to chemical reactions, the mass of each element is conserved. Consider a_{ij} as the number of atoms of element j in a molecule of species i and W_j the molecular weight of such atoms. The mass of all atoms j is given by

$$m_j = \sum_{i=1}^{n} \frac{a_{ij} W_j}{W_i} m_i, \tag{4.15}$$

where n is the number of species. The mass fraction of the element j is then written as

$$Z_j = \frac{m_j}{m} = \sum_{i=1}^{n} \frac{a_{ij} W_j}{W_i} Y_i, \tag{4.16}$$

where $j = 1, 2, \ldots, n_e$, and n_e is the total number of elements in the system. Note that $Y_i = m_i/m$ and $m = \sum_{i=1}^{n} m_i$ is the total mass of all the molecules in the mixture.

Adding the mass fractions in Eq. (3.77) yields the equation

$$\rho \frac{\partial Z_j}{\partial t} + \rho \vec{v} \cdot \vec{\nabla} Z_j = -\vec{\nabla} \cdot \left(\sum_{i=1}^{n} \frac{a_{ij} W_j}{W_i} \vec{j}_i \right), \tag{4.17}$$

where the chemical source term does not appear, because

$$\begin{aligned} \sum_{i=1}^{n} \frac{a_{ij} W_j}{W_i} \dot{w}_i &= \sum_{i=1}^{n} \left(\frac{a_{ij} W_j}{W_i} W_i \sum_{k=1}^{r} \nu_{ik} w_k \right) \\ &= W_j \sum_{i=1}^{n} \sum_{k=1}^{r} a_{ij} \nu_{ik} w_k = W_j \sum_{k=1}^{r} w_k \sum_{i=1}^{n} a_{ij} \nu_{ik} = 0 \end{aligned} \tag{4.18}$$

for each element j in any k reaction. Thus, the mass fraction of the element is preserved during combustion. Substituting the expression for the mass flow in Eq. (4.17), and assuming that all diffusivities are equal, $D_{ij} = D$, the balance equation for the mass fraction of elements takes the form

$$\rho \frac{\partial Z_j}{\partial t} + \rho \vec{v} \cdot \vec{\nabla} Z_j = \vec{\nabla} \cdot (\rho D \vec{\nabla} Z_j). \tag{4.19}$$

A similar equation can be obtained for the mixture fraction [2]. Because Z is defined by Eq. (4.7) as the mass fraction of the fuel stream into the mixture, Z is the sum of the mass fractions of the elements contained in the fuel stream,

$$Y_{\mathrm{F},u} = \sum_{j=1}^{n_e} Z_j. \tag{4.20}$$

With Eq. (4.8), the mixture fraction can be expressed as a sum of the mass fractions of the elements,

$$Z = \frac{\sum_{j=1}^{n_e} Z_j}{Y_{\mathrm{F},1}}. \tag{4.21}$$

Thus, assuming that all diffusivities are equal to D, the sum in Eq. (4.19) leads to the balance equation for the mixture fraction [2]

$$\rho \frac{\partial Z}{\partial t} + \rho \vec{v} \cdot \vec{\nabla} Z = \vec{\nabla} \cdot (\rho D \vec{\nabla} Z). \tag{4.22}$$

This equation has boundary conditions $Z = 1$ in the fuel stream and $Z = 0$ in the oxidant stream. The diffusion coefficient D is arbitrary, but because the maximum temperature determines the location of the reaction zone, the enthalpy diffusion is the most important transport process in the mixture fraction space. Therefore, the thermal diffusivity is a good choice to be the diffusion coefficient in (4.22).

In the case of mechanism with more than one step, the mixture fraction can be decomposed into n components, so that [3]

$$Z_1 + Z_2 + \cdots + Z_n = Z. \tag{4.23}$$

As an example, for a two-step mechanism,

$$\mathrm{I}\ \mathrm{F} + \mathrm{O_2} \rightarrow \mathrm{CO} + \text{Other products, such as } \mathrm{H_2O} \text{ and soot}, \tag{4.24}$$

$$\mathrm{II}\ \mathrm{CO} + \frac{1}{2}\mathrm{O_2} \rightarrow \mathrm{CO_2}, \tag{4.25}$$

the transport equations for the mass fractions of fuel, CO, and CO_2 are written as

$$\rho \frac{\partial Y_\mathrm{F}}{\partial t} + \rho \vec{v} \cdot \vec{\nabla} Y_\mathrm{F} = \vec{\nabla} \cdot (\rho D \vec{\nabla} Y_\mathrm{F}) + \dot{w}_{\mathrm{F,I}}, \tag{4.26}$$

$$\rho \frac{\partial Y_{CO}}{\partial t} + \rho \vec{v} \cdot \vec{\nabla} Y_{CO} = \vec{\nabla} \cdot (\rho D \vec{\nabla} Y_{CO}) + \dot{w}_{CO,I} + \dot{w}_{CO,II}, \tag{4.27}$$

$$\rho \frac{\partial Y_{CO_2}}{\partial t} + \rho \vec{v} \cdot \vec{\nabla} Y_{CO_2} = \vec{\nabla} \cdot (\rho D \vec{\nabla} Y_{CO_2}) + \dot{w}_{CO_2,II}, \tag{4.28}$$

where the subscript I and II refer to the reactions given in Eqs. (4.24) and (4.25), respectively. The production and consumption rates for the species in steps I and II are related by the equations

$$\frac{\dot{w}_{F,I}}{W_F} = -\frac{\dot{w}_{CO,I}}{x W_{CO}}, \tag{4.29}$$

$$\frac{\dot{w}_{CO,II}}{W_{CO}} = -\frac{\dot{w}_{CO_2,II}}{W_{CO_2}}, \tag{4.30}$$

where x are the moles of CO formed per mole of fuel burned. Using the linear combination

$$\dot{w}_F + \dot{w}_{CO} + \dot{w}_{CO_2} = 0, \tag{4.31}$$

it follows that

$$\frac{1}{\nu_F}\frac{d[F]}{dt} + \frac{1}{\nu_{CO}}\frac{d[CO]}{dt} + \frac{1}{\nu_{CO_2}}\frac{d[CO_2]}{dt} = 0 \Rightarrow \tag{4.32}$$

$$\frac{d\left(\frac{\rho Y_F}{W_F}\right)}{dt} + \frac{1}{x}\frac{d\left(\frac{\rho Y_{CO}}{W_{CO}}\right)}{dt} + \frac{1}{x}\frac{d\left(\frac{\rho Y_{CO_2}}{W_{CO_2}}\right)}{dt} = 0 \Rightarrow \tag{4.33}$$

$$\frac{dY_F}{W_F} + \frac{dY_{CO}}{x W_{CO}} + \frac{dY_{CO_2}}{x W_{CO_2}} = 0. \tag{4.34}$$

For the mechanism given by Eqs. (4.24) and (4.25), $\nu_{CO} = \nu_{CO_2} = x$. Integrating Eq. (4.34) from an unburned state (subscript u) to any state yields

$$\frac{Y_F - Y_{F,u}}{W_F} + \frac{Y_{CO} - Y_{CO,u}}{x W_{CO}} + \frac{Y_{CO_2} - Y_{CO_2,u}}{x W_{CO_2}} = 0. \tag{4.35}$$

Using the fact that $Y_{F,u} = Y_{F,1} Z$ and $Y_{CO,u} = Y_{CO_2,u} = 0$, it follows that

$$Z = \frac{1}{Y_{F,1}}\left(Y_F + \frac{W_F}{x W_{CO}} Y_{CO} + \frac{W_F}{x W_{CO_2}} Y_{CO_2}\right), \tag{4.36}$$

and therefore, Z_1, Z_2, and Z_3 are written as

$$Z_1 = \frac{Y_F}{Y_{F,1}}, \tag{4.37}$$

$$Z_2 = \frac{W_F}{x Y_{F,1} W_{CO}} Y_{CO}, \tag{4.38}$$

$$Z_3 = \frac{W_F}{x Y_{F,1} W_{CO_2}} Y_{CO_2}. \tag{4.39}$$

Substituting Eqs. (4.29), (4.30), (4.37)–(4.39) in Eqs. (4.26)–(4.28), one obtains:

$$\rho\frac{\partial Z_1}{\partial t}+\rho\vec{v}\cdot\vec{\nabla}Z_1=\vec{\nabla}\cdot(\rho D\vec{\nabla}Z_1)-\frac{W_F}{xY_{F,1}W_{CO}}\dot{w}_{CO,I}, \tag{4.40}$$

$$\rho\frac{\partial Z_2}{\partial t}+\rho\vec{v}\cdot\vec{\nabla}Z_2=\vec{\nabla}\cdot(\rho D\vec{\nabla}Z_2)+\frac{W_F}{xY_{F,1}W_{CO}}\left(\dot{w}_{CO,I}+\dot{w}_{CO,II}\right), \tag{4.41}$$

$$\rho\frac{\partial Z_3}{\partial t}+\rho\vec{v}\cdot\vec{\nabla}Z_3=\vec{\nabla}\cdot(\rho D\vec{\nabla}Z_3)-\frac{W_F}{xY_{F,1}W_{CO}}\dot{w}_{CO,II}. \tag{4.42}$$

Adding Eqs. (4.40)–(4.42) yields the equation of mixture fraction, Eq. (4.22).

Assuming that the other species of the mechanism are functions of CO and CO_2, the three amounts, Z_1, Z_2, and Z_3, can be used to determine the mass fractions of all species, which reduces the number of transport equations to be solved.

4.2 TURBULENT FLOWS

Most of the flow regimes that occur in nature are turbulent. Some examples are the atmosphere boundary layer, except in cases of very stable atmosphere, the flow of natural gas and oil in pipelines, the flow of water in rivers and canals, and the flow in pumps and turbines.

The study of turbulence is clearly an interdisciplinary activity that has a wide range of applications. In fluid dynamics, laminar flow is the exception, not the rule [4].

Hinze [5] defines turbulence as an irregular condition of flow in which the various quantities show a random variation involved with the coordinates of time and space, which can be differentiated statistically from their mean values.

A turbulent flow has some important characteristics, such as:

1. *Irregularity*: A turbulent flow is chaotic. Various quantities such as velocity, temperature, and pressure fluctuate.
2. *Diffusivity*: The turbulence has a great capacity for mixing. The turbulent flow increases the rate of mass transfer, heat, and momentum. This characteristic is important in many applications.
3. *High Reynolds number*: Turbulent flow originates generally from an instability in a laminar flow when the Reynolds number is sufficiently high. This instability is related to interactions between the viscous terms and the nonlinear inertia terms from the momentum equations;
4. *Three-dimensional fluctuations*: Turbulence is rotational and three dimensional, showing high levels of vorticity.
5. *Dissipation*: The turbulent flow requires a continuous supply of energy, otherwise the turbulence decays rapidly as the kinetic energy is converted into internal energy by viscous shear stresses.

6. *Continuous phenomenon*: Turbulence is a continuous phenomenon, governed by the equations of fluid mechanics. Even the smallest scales of turbulence occurring in a turbulent flow are larger than the scales of molecular size.

4.2.1 Scales of Turbulence

The combustion, even without turbulence, and geochemistry involve large variations in time and length scales [6]. The turbulent combustion is a result of the interaction between chemistry and turbulence.

The turbulent velocity field can be represented by eddies of different sizes. An eddy can be described as a turbulent motion located within a region of size l, which shows a moderately coherent structure in this region [7]. Coherent structure is understood as a region of space that, at a given time, has some sort of organization in relation to any variable related to the flow (velocity, vorticity, pressure, density, temperature, etc.).

The scales of turbulence represent the magnitude of the variables involved in the flow. The principal scales of turbulence are related to length, time, speed, energy, and vorticity. Kolmogorov defined the significant smaller scales that can occur in a turbulent flow based on the hypothesis that the larger eddies transfer energy to smaller eddies, and these, in turn, transfer energy to even smaller eddies. This process results in an energy transfer in a cascade form, from larger eddies to smaller eddies, until the energy of smallest eddies is dissipated as heat by viscous forces.

Considering a vortex of characteristic size r and characteristic speed v_r, the Reynolds number is defined by

$$Re_r = \frac{v_r r}{\nu}, \tag{4.43}$$

where ν corresponds to the kinematic viscosity of the fluid.

The rate of viscous dissipation ϵ is an important parameter that represents the energy injected from large scales to small scales of turbulence. This rate can be described by the relation [4]

$$\epsilon = \frac{A v_r^3}{r}, \tag{4.44}$$

where A corresponds to an empirical parameter, whose value is of order 1 [8]. Substituting this relation in Eq. (4.43) yields the following expression for the turbulent Reynolds number depending on ϵ

$$Re_r = \frac{\left(\epsilon r^4\right)^{1/3}}{\nu}. \tag{4.45}$$

Considering that, for the scale r analyzed, the viscous effects are small, then it is clear that $Re_r > 1$. If dimension r decreases, the number of Re_r also decreases. So there is a limit size l_η, where, for scales of length smaller than this value, the Reynolds number becomes smaller than 1, and viscous effects start to dominate over inertial effects. The vortices with sizes less than l_η are dissipated by viscous effects and fail to develop. Therefore, this scale l_η corresponds to the Kolmogorov length scale and can be written according to the equation

$$l_\eta = \left(\frac{\nu^3}{\epsilon}\right)^{1/4}. \tag{4.46}$$

Performing a dimensional analysis and expressing the characteristic time as a function of ϵ and ν, one obtains the Kolmogorov time scale given by

$$\tau_\eta = \left(\frac{\nu}{\epsilon}\right)^{1/2}. \tag{4.47}$$

Using the ratio of length scale (4.46) and time scale (4.47), results in the following expression for the speed in the Kolmogorov velocity scale:

$$v_\eta = (\nu\epsilon)^{1/4}. \tag{4.48}$$

The vorticity of a fluid corresponds to a vector field given by the velocity curl. Using Eq. (4.47) with timescale yields the Kolmogorov vorticity scale

$$\omega_\eta = \left(\frac{\epsilon}{\nu}\right)^{1/2}. \tag{4.49}$$

The largest scales of a flow are determined by the geometry that gave rise to them. Considering L the scale of length representative of the structure of the flow and U a representative velocity scale, the scales of time, vorticity, and energy are given by the following relations:

$$t = L/U, \tag{4.50}$$

$$\omega = U/L, \tag{4.51}$$

$$E = U^2. \tag{4.52}$$

4.2.2 Reynolds and Favre Averages

To express the turbulent flow, usually the dependent variables are expressed as the sum of an average and a fluctuation in order to simplify the equations system. The most commonly used averages are those of Reynolds and Favre.

The Reynolds average decomposes the variable in an average value $\overline{\Phi}$ and a fluctuation Φ', according to the relations:

$$\Phi = \overline{\Phi} + \Phi', \qquad \text{where} \quad \overline{\Phi'} = 0, \tag{4.53}$$

$$\overline{\Phi} = \lim_{\Delta t \to \infty}\left(\frac{1}{\Delta t}\int_{t_0}^{t_0+\Delta t} \Phi(t)\,\mathrm{d}t\right) \tag{4.54}$$

and the decomposition has the following properties

$$
\begin{aligned}
&\overline{\overline{\Phi}} = \overline{\Phi},\\
&\overline{\Phi'} = 0,\\
&\overline{\Phi \cdot \overline{\chi}} = \overline{\left(\overline{\Phi} + \Phi'\right) \cdot \overline{\chi}} = \overline{\Phi} \cdot \overline{\chi},\\
&\overline{\Phi' \cdot \overline{\chi}} = 0\\
&\overline{\Phi + \chi} = \overline{\left(\overline{\Phi} + \Phi'\right) + (\overline{\chi} + \chi')} = \overline{\Phi} + \overline{\chi},\\
&\overline{\frac{\partial \Phi}{\partial x}} = \frac{\partial \overline{\left(\overline{\Phi} + \Phi'\right)}}{\partial x} = \frac{\partial \overline{\Phi}}{\partial x},\\
&\overline{\int \Phi \, \mathrm{d}x} = \int \overline{\left(\overline{\Phi} + \Phi\right)'} \mathrm{d}x = \int \overline{\Phi} \, \mathrm{d}x.
\end{aligned}
\tag{4.55}
$$

The Favre average decomposes the dependent variable in an average value $\tilde{\Phi}$ and a fluctuation Φ'' too, and this weighted average in density is given by:

$$\Phi = \tilde{\Phi} + \Phi'', \qquad \text{where } \overline{\rho \, \Phi''} = 0, \tag{4.56}$$

$$\tilde{\Phi} = \frac{\overline{\rho\Phi}}{\overline{\rho}} = \lim_{\Delta t \to \infty} \left[\frac{\int_t^{t+\Delta t} \rho \Phi \, \mathrm{d}t}{\int_t^{t+\Delta t} \rho \, \mathrm{d}t} \right]. \tag{4.57}$$

Similar to the Reynolds average, the Favre average has the properties:

$$
\begin{aligned}
&\overline{\rho\Phi} = \overline{\rho}\tilde{\Phi},\\
&\overline{\rho\Phi''} = 0,\\
&\overline{\rho\tilde{\Phi}} = \overline{\rho}\tilde{\Phi}.
\end{aligned}
\tag{4.58}
$$

In turbulent regimes, where there are large fluctuations in density, the average of Favre is preferred, in order to avoid the appearance of terms containing explicitly products of fluctuations, such as $(\rho' v_j')$.

The following equations of continuity, momentum, energy, and mixture fraction are presented in terms of Favre averages.

Continuity Equation

Applying the Reynolds average to the continuity equation (3.8) yields the expression

$$\overline{\frac{\partial \rho}{\partial t} + \frac{\partial \left(\rho v_j\right)}{\partial x_j}} = 0. \tag{4.59}$$

Substituting the relation given by (4.56) in (4.59), one obtains

$$\frac{\partial \overline{\rho}}{\partial t} + \frac{\partial \left[\overline{\rho\left(\tilde{v}_j + v_j''\right)}\right]}{\partial x_j} = 0. \tag{4.60}$$

Using the relation $\overline{\rho\,\Phi''} = 0$ yields the equation of continuity in the Favre average

$$\frac{\partial \overline{\rho}}{\partial t} + \frac{\partial \left(\overline{\rho}\tilde{v}_j\right)}{\partial x_j} = 0. \tag{4.61}$$

Equation of Momentum

Using the Reynolds average in the conservation of momentum gives

$$\overline{\rho\left(\frac{\partial v_i}{\partial t} + v_j\frac{\partial v_i}{\partial x_j}\right)} = \overline{-\frac{\partial p}{\partial x_i} + \frac{\partial \tau_{ij}}{\partial x_j}} + \rho g_i. \tag{4.62}$$

Decomposing the velocity in terms of Favre averages results in

$$\overline{\rho\frac{\partial\left(\tilde{v}_i + v_i''\right)}{\partial t}} + \overline{\rho\left(\tilde{v}_j + v_j''\right)\frac{\partial\left(\tilde{v}_i + v_i''\right)}{\partial x_j}} = -\frac{\partial \overline{p}}{\partial x_i} + \frac{\partial \overline{\tau_{ij}}}{\partial x_j} + \overline{\rho} g_i. \tag{4.63}$$

Inserting the following relations

$$\overline{\rho\frac{\partial v_i''}{\partial t}} = \frac{\partial \overline{\left(\rho v_i''\right)}}{\partial t} - \overline{v_i''\frac{\partial \rho}{\partial t}} = -\overline{v_i''\frac{\partial \rho}{\partial t}}, \tag{4.64}$$

$$\overline{\rho\tilde{v}_j\frac{\partial v_i''}{\partial x_j}} = \tilde{v}_j\frac{\partial \overline{\left(\rho v_i''\right)}}{\partial x_j} - \overline{\tilde{v}_j v_i''\frac{\partial \rho}{\partial x_j}} = -\overline{\tilde{v}_j v_i''\frac{\partial \rho}{\partial x_j}}, \tag{4.65}$$

$$\overline{\rho v_j''\frac{\partial v_i''}{\partial x_j}} = \frac{\partial \overline{\left(\rho v_j'' v_i''\right)}}{\partial x_j} - \overline{v_i''\frac{\partial \left(\rho v_j''\right)}{\partial x_j}}, \tag{4.66}$$

$$\overline{\rho v_j'' v_i''} = \overline{\rho}\widetilde{v_j'' v_i''}, \tag{4.67}$$

in Eq. (4.63) yields the equation

$$\begin{aligned}\overline{\rho}\frac{\partial \tilde{v}_i}{\partial t} + \overline{\rho}\tilde{v}_j\frac{\partial \tilde{v}_i}{\partial x_j} + \frac{\partial\left(\overline{\rho}\widetilde{v_j'' v_i''}\right)}{\partial x_j} - \overline{v_i''\left[\frac{\partial \rho}{\partial t} + \frac{\partial\left(\rho v_j''\right)}{\partial x_j} + \tilde{v}_j\frac{\partial \rho}{\partial x_j}\right]} \\ = -\frac{\partial \overline{p}}{\partial x_i} + \frac{\partial \overline{\tau_{ij}}}{\partial x_j} + \overline{\rho} g_i.\end{aligned} \tag{4.68}$$

Using the continuity equation yields

$$\frac{\partial \rho}{\partial t} + \frac{\partial\left(\rho v_j''\right)}{\partial x_j} + \tilde{v}_j\frac{\partial \rho}{\partial x_j} = -\rho\frac{\partial \tilde{v}_j}{\partial x_j}. \tag{4.69}$$

Substituting Eq. (4.69) in Eq. (4.68), the fourth term of Eq. (4.68), gives $\overline{\rho v_i''}\frac{\partial \tilde{v}_j}{\partial x_j}$, whose value is zero, yielding the equation of conservation of momentum in the Favre form

$$\overline{\rho}\frac{\partial \widetilde{v}_i}{\partial t} + \overline{\rho}\widetilde{v}_j\frac{\partial \widetilde{v}_i}{\partial x_j} = -\frac{\partial \overline{p}}{\partial x_i} + \frac{\partial \overline{\tau_{ij}}}{\partial x_j} - \frac{\partial\left(\overline{\rho}\widetilde{v_j'' v_i''}\right)}{\partial x_j} + \overline{\rho}g_i. \tag{4.70}$$

The term $\overline{\rho}\widetilde{v_j'' v_i''}$ is called the Reynolds tensor and corresponds to the stress exerted by the turbulent fluctuations of the fluid.

Equation of Energy in Terms of Enthalpy

First, the Reynolds average is applied to the energy equation in terms of enthalpy (3.57), disregarding the term $\dot{q}_v$, and yielding

$$\overline{\rho\left(\frac{\partial h}{\partial t} + v_j\frac{\partial h}{\partial x_j}\right)} = \overline{-\frac{\partial \dot{q}}{\partial x_j} + \frac{\partial p}{\partial t} + v_j\frac{\partial p}{\partial x_j} + \tau_{ij}\frac{\partial v_i}{\partial x_j}}. \tag{4.71}$$

Considering that the enthalpy is the sum of a time average ($\tilde{h}$) and a fluctuation (h''), and similarly for velocity v_j, which corresponds to the sum of $\widetilde{v}_j$ and v_j'', it follows that

$$\begin{aligned}&\overline{\rho\frac{\partial\left(\tilde{h}+h''\right)}{\partial t}} + \overline{\rho\left(\widetilde{v}_j + v_j''\right)\frac{\partial\left(\tilde{h}+h''\right)}{\partial x_j}}\\&= -\frac{\partial \bar{q}}{\partial x} + \frac{\partial \overline{p}}{\partial t} + \overline{\left(\tilde{v}_j + v_j''\right)\frac{\partial p}{\partial x_j}} + \overline{\tau_{ij}\frac{\partial v_i}{\partial x_j}}.\end{aligned} \tag{4.72}$$

Rearranging terms of this equation, one obtains

$$\begin{aligned}&\overline{\rho}\frac{\partial \tilde{h}}{\partial t} + \overline{\rho}\widetilde{v}_j\frac{\partial \tilde{h}}{\partial x_j} + \overline{\rho\frac{\partial h''}{\partial t}} + \overline{\rho\widetilde{v}_j\frac{\partial h''}{\partial x_j}} + \overline{\rho v_j''\frac{\partial h''}{\partial x_j}}\\&= -\frac{\partial \bar{\dot{q}}}{\partial x} + \frac{\partial \overline{p}}{\partial t} + \tilde{v}_j\frac{\partial \overline{p}}{\partial x_j} + \overline{v_j''\frac{\partial p}{\partial x_j}} + \overline{\tau_{ij}\frac{\partial v_i}{\partial x_j}}.\end{aligned} \tag{4.73}$$

Because

$$\overline{\rho\frac{\partial h''}{\partial t}} = \frac{\partial \overline{\rho h''}}{\partial t} - \overline{h''\frac{\partial \rho}{\partial t}} = -\overline{h''\frac{\partial \rho}{\partial t}}, \tag{4.74}$$

$$\overline{\rho\tilde{v}_j\frac{\partial h''}{\partial x_j}} = \frac{\partial}{\partial x_j}\left(\tilde{v}_j\overline{\rho h''}\right) - \overline{h''\frac{\partial}{\partial x_j}\left(\rho\tilde{v}_j\right)} = -\overline{h''\frac{\partial}{\partial x_j}\left(\rho\tilde{v}_j\right)}, \tag{4.75}$$

$$\overline{\rho v_j''\frac{\partial h''}{\partial x_j}} = \frac{\partial}{\partial x_j}\left(\overline{\rho}\widetilde{v_j'' h''}\right) - \overline{h''\frac{\partial\left(\rho v_j''\right)}{\partial x_j}}, \tag{4.76}$$

it follows that

$$\overline{\rho\frac{\partial h''}{\partial t}} + \overline{\rho\tilde{v}_j\frac{\partial h''}{\partial x_j}} + \overline{\rho v_j''\frac{\partial h''}{\partial x_j}}$$
$$= \overline{-h''\left[\frac{\partial \rho}{\partial t} + \frac{\partial}{\partial x_j}\left(\tilde{v}_j + v_j''\right)\right]} + \frac{\partial}{\partial x_j}\left(\bar{\rho}\widetilde{v_j''h''}\right). \quad (4.77)$$

Substituting Eq. (4.77) in (4.73) yields the following expression for the Favre averaged energy conservation

$$\bar{\rho}\frac{D\tilde{h}}{Dt} = \frac{\partial \bar{p}}{\partial t} + \tilde{v}_j\frac{\partial \bar{p}}{\partial x_j} + \overline{v_j''\frac{\partial p}{\partial x_j}} - \frac{\partial \bar{\dot{q}}}{\partial x} + \overline{\tau_{ij}\frac{\partial v_i}{\partial x_j}} - \frac{\partial}{\partial x_j}\left(\bar{\rho}\widetilde{v_j''h''}\right). \quad (4.78)$$

Equation of Energy in Terms of Temperature

Applying the Reynolds average to the conservation of energy expressed in terms of temperature gives the expression

$$\overline{\rho c_p\left(\frac{\partial T}{\partial t} + v_j\frac{\partial T}{\partial x_j}\right)} = \overline{\frac{\partial \dot{q}}{\partial x_j} + \frac{\partial p}{\partial t} - \sum_{i=1}^{n} h_i\dot{w}_i + q_R}. \quad (4.79)$$

Using the Favre average on this equation gives

$$\overline{\rho c_p\frac{\partial\left(\tilde{T}+T''\right)}{\partial t}} + \overline{\rho c_p\left(\tilde{v}_j + v_j''\right)\frac{\partial\left(\tilde{T}+T''\right)}{\partial x_j}}$$
$$= \frac{\overline{\partial \dot{q}}}{\partial x_j} + \frac{\partial \bar{p}}{\partial t} - \sum_{i=1}^{n}\overline{h_i\dot{w}_i} + \overline{q_R}. \quad (4.80)$$

Considering the specific heat capacity constant and using the following relations

$$\overline{\rho\frac{\partial T''}{\partial t}} = \frac{\partial\left(\overline{\rho T''}\right)}{\partial t} - \overline{T''\frac{\partial \rho}{\partial t}} = -\overline{T''\frac{\partial \rho}{\partial t}}, \quad (4.81)$$

$$\overline{\rho\tilde{v}_j\frac{\partial T''}{\partial x_j}} = \tilde{v}_j\frac{\partial\left(\overline{\rho T''}\right)}{\partial x_j} - \overline{\tilde{v}_jT''\frac{\partial \rho}{\partial x_j}} = -\overline{\tilde{v}_jT''\frac{\partial \rho}{\partial x_j}}, \quad (4.82)$$

$$\overline{\rho v_j''\frac{\partial T''}{\partial x_j}} = \frac{\partial\left(\overline{\rho v_j''T''}\right)}{\partial x_j} - \overline{T''\frac{\partial\left(\rho v_j''\right)}{\partial x_j}}, \quad (4.83)$$

$$\overline{\rho v_j''T''} = \bar{\rho}\widetilde{v_j''T''}, \quad (4.84)$$

in Eq. (4.80) yields

$$\overline{\rho}c_p\left(\frac{\partial\tilde{T}}{\partial t}+\widetilde{v_j}\frac{\partial\tilde{T}}{\partial x_j}\right)+c_p\frac{\partial\left(\overline{\rho}\widetilde{v_j^{''}T^{''}}\right)}{\partial x_j}-\overline{c_pT^{''}\left(\frac{\partial\rho}{\partial t}+v_j^{''}\frac{\partial\rho}{\partial x_j}+\frac{\partial\left(\rho v_j^{''}\right)}{\partial x_j}\right)}$$
$$=\frac{\partial\overline{\dot{q}}}{\partial x_j}+\frac{\partial\overline{p}}{\partial t}-\sum_{i=1}^{n}\overline{h_i\dot{w}_i}+\overline{q}_R. \tag{4.85}$$

Replacing Eq. (4.69) in (4.85) results in the Favre averaged energy equation for temperature

$$\overline{\rho}c_p\left(\frac{\partial\tilde{T}}{\partial t}+\widetilde{v_j}\frac{\partial\tilde{T}}{\partial x_j}\right)=\frac{\partial\overline{\dot{q}}}{\partial x_j}-c_p\frac{\partial\left(\overline{\rho}\widetilde{v_j^{''}T^{''}}\right)}{\partial x_j}+\frac{\partial\overline{p}}{\partial t}-\sum_{i=1}^{n}\overline{h_i\dot{w}_i}+\overline{q}_R. \tag{4.86}$$

Equations for Turbulent Kinetic Energy

Consider the expression for the conservation of momentum, given by Eq. (3.32). Multiplying this equation by velocity $\vec{v}$ yields the following equation, written in Cartesian coordinates, in the directions i and k

$$\rho\frac{\partial v_i}{\partial t}v_k+\rho v_j\frac{\partial v_i}{\partial x_j}v_k=-\frac{\partial p}{\partial x_i}v_k+\frac{\partial\tau_{ij}}{\partial x_j}v_k+\rho v_k g_i, \tag{4.87}$$

which can be rewritten as

$$\frac{\partial}{\partial t}(\rho v_k v_i)+\frac{\partial}{\partial x_j}\left(\rho v_j v_k v_i\right)$$
$$=v_i\left(\rho\frac{\partial v_k}{\partial t}+\rho v_j\frac{\partial v_k}{\partial x_j}\right)-v_k\frac{\partial p}{\partial x_i}+v_k\frac{\partial\tau_{ij}}{\partial x_j}+\rho v_k g_i. \tag{4.88}$$

The term in parentheses on the right-hand side of Eq. (4.88) can be written as a function of the momentum equation in the direction k, so that the equation can be written as

$$\frac{\partial}{\partial t}(\rho v_k v_i)+\frac{\partial}{\partial x_j}\left(\rho v_j v_k v_i\right)=-v_i\frac{\partial p}{\partial k}-v_k\frac{\partial p}{\partial x_i}+v_i\frac{\partial\tau_{kj}}{\partial x_j}+v_k\frac{\partial\tau_{ij}}{\partial x_j}$$
$$+\rho v_i g_k+\rho v_k g_i. \tag{4.89}$$

After applying the Favre decomposition to the components of the velocity vector and the Reynolds decomposition to the shear stress, one obtains

$$\frac{\partial}{\partial t}\left(\overline{\rho}\tilde{v}_k\tilde{v}_i+\overline{\rho}\widetilde{v_k^{''}v_i^{''}}\right)+\frac{\partial}{\partial x_j}\left(\overline{\rho}\tilde{v}_j\tilde{v}_k\tilde{v}_i+\overline{\rho}\tilde{v}_i\widetilde{v_j^{''}v_k^{''}}+\overline{\rho}\tilde{v}_j\widetilde{v_i^{''}v_k^{''}}+\overline{\rho}\tilde{v}_k\widetilde{v_j^{''}v_i^{''}}\right.$$
$$\left.+\overline{\rho}\widetilde{v_j^{''}v_k^{''}v_i^{''}}\right)=-\tilde{v}_i\frac{\partial\overline{p}}{\partial x_k}-\tilde{v}_k\frac{\partial\overline{p}}{\partial x_i}-\overline{v_i^{''}\frac{\partial p}{\partial x_k}}-\overline{v_k^{''}\frac{\partial p}{\partial x_i}}$$
$$+\tilde{v}_i\frac{\partial\overline{\tau_{kj}}}{\partial x_j}+\tilde{v}_k\frac{\partial\overline{\tau_{ij}}}{\partial x_j}+\overline{v_i^{''}}\frac{\partial\overline{\tau_{kj}}}{\partial x_j}+\overline{v_k^{''}}\frac{\partial\overline{\tau_{ij}}}{\partial x_j}+\overline{v_i^{''}\frac{\partial\tau_{kj}^{'}}{\partial x_j}}$$
$$+\overline{v_k^{''}\frac{\partial\tau_{ij}^{'}}{\partial x_j}}+\overline{\rho}\tilde{v}_i g_k+\overline{\rho}\tilde{v}_k g_i. \tag{4.90}$$

Multiplying the equation of momentum (4.70) by the average velocity vector yields the following equation written in Cartesian coordinates:

$$\overline{\rho}\frac{\partial\tilde{v}_i}{\partial t}\tilde{v}_k+\overline{\rho}\tilde{v}_j\frac{\partial\tilde{v}_i}{\partial x_j}\tilde{v}_k=-\frac{\partial\overline{p}}{\partial x_i}\tilde{v}_k+\frac{\partial\overline{\tau_{ij}}}{\partial x_j}\tilde{v}_k-\frac{\partial\left(\overline{\rho}\widetilde{v_j^{''}v_i^{''}}\right)}{\partial x_j}\tilde{v}_k+\overline{\rho}\tilde{v}_k g_i. \tag{4.91}$$

Working this equation mathematically, similarly as with Eq. (4.88), one obtains

$$\frac{\partial}{\partial t}\left(\overline{\rho}\tilde{v}_k\tilde{v}_i\right)+\frac{\partial}{\partial x_j}\left(\overline{\rho}\tilde{v}_k\tilde{v}_j\tilde{v}_i\right)=-\overline{v}_i\frac{\partial\overline{p}}{\partial x_k}-\tilde{v}_k\frac{\partial\overline{p}}{\partial x_i}+\overline{v}_i\frac{\partial\overline{\tau_{kj}}}{\partial x_j}+\overline{v}_k\frac{\partial\overline{\tau_{ij}}}{\partial x_j}+$$
$$-\overline{v}_i\frac{\partial}{\partial x_j}\left(\overline{\rho}\widetilde{v_j^{''}v_k^{''}}\right)-\overline{v}_k\frac{\partial}{\partial x_j}\left(\overline{\rho}\widetilde{v_j^{''}v_i^{''}}\right)+\overline{\rho}\tilde{v}_i g_k+\overline{\rho}\tilde{v}_k g_i. \tag{4.92}$$

Subtracting the expressions (4.90) and (4.92) gives an expression for the Reynolds tensor given by equation

$$\frac{\partial}{\partial t}\left(\overline{\rho}\widetilde{v_k^{''}v_i^{''}}\right)+\frac{\partial}{\partial x_j}\left(\tilde{v}_j\overline{\rho}\widetilde{v_k^{''}v_i^{''}}\right)+\frac{\partial}{\partial x_j}\left(\overline{\rho}\widetilde{v_j^{''}v_k^{''}v_i^{''}}\right)=-\overline{v_i^{''}\frac{\partial p}{\partial x_k}}-\overline{v_k^{''}\frac{\partial p}{\partial x_i}}$$
$$+\overline{v_i^{''}\frac{\partial\tau_{kj}^{'}}{\partial x_j}}+\overline{v_k^{''}\frac{\partial\tau_{ij}^{'}}{\partial x_j}}+\overline{v_i^{''}}\frac{\partial\overline{\tau_{kj}}}{\partial x_j}+\overline{v_k^{''}}\frac{\partial\overline{\tau_{ij}}}{\partial x_j}-\overline{\rho}\widetilde{v_j^{''}v_k^{''}}\frac{\partial\tilde{v}_i}{\partial x_j}-\overline{\rho}\widetilde{v_j^{''}v_i^{''}}\frac{\partial\tilde{v}_k}{\partial x_j}. \tag{4.93}$$

In the case $i=k$, this expression represents the fluctuation of the flow kinetic energy, and the equation takes the form

$$\frac{\partial}{\partial t}\left(\frac{1}{2}\overline{\rho}\widetilde{v_i^{''2}}\right)+\frac{\partial}{\partial x_j}\left(\tilde{v}_j\frac{1}{2}\overline{\rho}\widetilde{v_i^{''2}}\right)+\frac{\partial}{\partial x_j}\left(\frac{1}{2}\overline{\rho}\widetilde{v_j^{''}v_i^{''2}}\right)=-\frac{\overline{\partial\left(v_i^{''}p\right)}}{\partial x_i}+\overline{p\frac{\partial v_i^{''}}{\partial x_i}}$$
$$+\overline{v_i^{''}\frac{\partial\tau_{ij}^{'}}{\partial x_j}}+\overline{v_i^{''}}\frac{\partial\overline{\tau_{ij}}}{\partial x_j}-\overline{\rho}\widetilde{v_j^{''}v_i^{''}}\frac{\partial\tilde{v}_i}{\partial x_j}. \tag{4.94}$$

Turbulent kinetic energy is defined as $\tilde{k}=\widetilde{v^{''2}}/2$. Substituting this relation into Eq. (4.94), and disregarding the terms $\overline{v_i^{''}}$ and $\overline{p\partial v_i^{''}/\partial x_i}$ yields the equation

$$\bar{\rho}\frac{\partial \tilde{k}}{\partial t} + \bar{\rho}\tilde{v}_j\frac{\partial \tilde{k}}{\partial x_j} = -\frac{\partial}{\partial x_j}\left(\frac{1}{2}\bar{\rho}\widetilde{v_j''v_i''^2} + \overline{v_i''p'}\delta_{i,j}\right) + \frac{\partial \overline{v_i''\tau_{ij}'}}{\partial x_j} - \overline{\tau_{ij}'\frac{\partial v_i''}{\partial x_j}} - \bar{\rho}\widetilde{v_j''v_i''}\frac{\partial \tilde{v}_i}{\partial x_j}. \tag{4.95}$$

Using the assumption that the fluid has Newtonian behavior, the fluctuation of Reynolds shear stress yields

$$\vec{\vec{\tau}}' = \mu\left[\vec{\nabla}\vec{v}' + \left(\vec{\nabla}\vec{v}'\right)^T - \frac{2}{3}\left(\vec{\nabla}\cdot\vec{v}'\right)\vec{\vec{I}}\right]. \tag{4.96}$$

For an incompressible flow, $v' \approx v''$, the fifth term of Eq. (4.95) can be approximated by

$$\frac{\partial \overline{v_i''\tau_{ij}'}}{\partial x_j} \approx \frac{\partial}{\partial x_j}\left[\bar{\rho}\nu\frac{\partial}{\partial x_j}\left(\frac{1}{2}\widetilde{v_i''^2}\right)\right]. \tag{4.97}$$

Substituting this expression into Eq. (4.95) results in

$$\rho\frac{D\tilde{k}}{Dt} = \frac{\partial}{\partial x_j}\left[\bar{\rho}\nu\frac{\partial \tilde{k}}{\partial x_j} - \left(\frac{1}{2}\bar{\rho}\widetilde{v_j''v_i''^2} + \overline{v_i''p'}\delta_{ij}\right)\right] - \bar{\rho}\widetilde{v_j''v_i''}\frac{\partial \tilde{v}_i}{\partial x_j} - \rho\tilde{\epsilon}, \tag{4.98}$$

where $\tilde{\epsilon}$ corresponds to the turbulent dissipation, defined as

$$\rho\tilde{\epsilon} = \overline{\tau_{ij}'\frac{\partial v_i''}{\partial x_j}}. \tag{4.99}$$

The term $\left(\frac{1}{2}\bar{\rho}\widetilde{v_j''v_i''^2} + \overline{v_i''p'}\delta_{ij}\right)$ can be modeled using the gradient hypothesis [7], resulting in

$$\left(\frac{1}{2}\bar{\rho}\widetilde{v_j''v_i''^2} + \overline{v_i''p'}\delta_{ij}\right) = -\bar{\rho}\frac{\nu_t}{S_{ck}}\frac{\partial \tilde{k}}{\partial x_j}. \tag{4.100}$$

Substituting this expression into Eq. (4.98) yields the differential equation for the turbulent kinetic energy, given by

$$\rho\frac{D\tilde{k}}{Dt} = \frac{\partial}{\partial x_j}\left[\bar{\rho}\left(\nu + \frac{\nu_t}{S_{ck}}\right)\frac{\partial \tilde{k}}{\partial x_j}\right] - \bar{\rho}\widetilde{v_j''v_i''}\frac{\partial \tilde{v}_i}{\partial x_j} - \rho\tilde{\epsilon}. \tag{4.101}$$

Writing the equation for turbulent kinetic energy in vector coordinates gives

$$\bar{\rho}\frac{D\tilde{k}}{Dt} = \vec{\nabla}\cdot\left[\bar{\rho}\left(\nu + \frac{\nu_t}{S_{ck}}\right)\vec{\nabla}\tilde{k}\right] - \bar{\rho}\widetilde{\vec{v}''\vec{v}''} : \left(\vec{\nabla}\tilde{\vec{v}}\right) - \bar{\rho}\tilde{\epsilon}. \tag{4.102}$$

In order to close the system of equations, hypotheses or models for the turbulent viscosity are required. Boussinesq [9] was one of the first to develop a model for the Reynolds tensor, introducing the concept of turbulent viscosity.

4.2.3 Turbulent Viscosity

Consider the expression for the momentum obtained in (4.70)

$$\bar{\rho}\frac{\partial \tilde{v}_i}{\partial t} + \bar{\rho}\tilde{v}_j\frac{\partial \tilde{v}_i}{\partial x_j} = -\frac{\partial \bar{p}}{\partial x_i} + \frac{\partial \overline{\tau_{ij}}}{\partial x_j} - \frac{\partial \left(\bar{\rho}\widetilde{v_j'' v_i''}\right)}{\partial x_j} + \bar{\rho}g_i, \tag{4.103}$$

where the term $\bar{\rho}\widetilde{v_j'' v_i''}$ is called the Reynolds tensor.

Boussinesq assumed that the Reynolds tensor has a behavior similar to the viscous stress, which is proportional to the gradient of velocity multiplied by a coefficient of proportionality, called the turbulent viscosity,

$$-\bar{\rho}\widetilde{u'' v''} = \bar{\rho}\nu_t\frac{\partial \tilde{u}}{\partial y}. \tag{4.104}$$

Generalizing this expression yields in Cartesian coordinates

$$\bar{\rho}\widetilde{v_j'' v_i''} = \begin{cases} -\bar{\rho}\nu_t\left[2\dfrac{\partial \tilde{v}_j}{\partial x_i} - \dfrac{2}{3}\dfrac{\partial \tilde{v}_k}{\partial x_k}\right] + \dfrac{2}{3}\bar{\rho}\tilde{k}, & i = j, \\ -\bar{\rho}\nu_t\left[\dfrac{\partial \tilde{v}_j}{\partial x_i} + \dfrac{\partial \tilde{v}_i}{\partial x_j}\right], & i \neq j, \end{cases} \tag{4.105}$$

where $\tilde{k}$ corresponds to the turbulent kinetic energy in the average of Favre. In vectorial coordinates, the following equation is obtained:

$$\bar{\rho}\widetilde{\vec{v}''\vec{v}''} = -\bar{\rho}\nu_t\left[\vec{\nabla}\vec{\tilde{v}} + \left(\vec{\nabla}\vec{\tilde{v}}\right)^T - \frac{2}{3}\left(\vec{\nabla}\cdot\vec{\tilde{v}}\right)\vec{\vec{I}}\right] + \frac{2}{3}\bar{\rho}\tilde{k}\vec{\vec{I}}. \tag{4.106}$$

Contrary to molecular viscosity, which corresponds to a property of the fluid, the turbulent viscosity is a characteristic variable of the flow. This parameter is a simplification that uses concepts of laminar flow in turbulent flow [10].

In Boussinesq's original concept, the eddy viscosity was assumed to be spatially constant. However, this assumption is valid only if the turbulent flow field is homogeneous, which is not frequent. For general cases, it is necessary to consider the variation of turbulent viscosity along the flow.

A simple alternative is the use of semi-empirical expressions, for example, Agrawal and Prasad's [11] expression for an axisymmetric turbulent jet, written in cylindrical coordinates as

$$\tilde{U}(r,z) = U_c(z)\exp(-\xi^2), \qquad \xi = \frac{r}{c\,z}, \tag{4.107}$$

where c corresponds to an empirical parameter, while z and r correspond to the axial and radial coordinates, respectively. The variable U_c corresponds to the axial velocity in the centerline of an axisymmetric jet, which varies proportionally with z^{-1} [12].

There are other models in the literature that will be discussed in the following sections.

(a) Prandtl's Model of One Equation

In models of one equation, the turbulent viscosity is expressed in terms of a single turbulent quantity. Some of the principal models of this category correspond to the models of Prandtl [13], Baldwin-Barth [14], and Spalart-Allmaras [15].

The Prandtl's model defines the eddy viscosity as a function of turbulent kinetic energy $\tilde{k}$ in accordance with the expression

$$\nu_t = c_D \frac{\tilde{k}^2}{\tilde{\epsilon}}, \tag{4.108}$$

and the turbulent dissipation $\widetilde{\epsilon}$ is written as a function of $\widetilde{k}$ according to

$$\tilde{\epsilon} = c_D \frac{\tilde{k}^{\frac{3}{2}}}{l}, \tag{4.109}$$

where the variable l corresponds to the turbulent length scale.

The differential equation for $\widetilde{k}$ is analogous to the expression employed in the $\widetilde{k} - \widetilde{\epsilon}$ model (to be discussed in the following), given by

$$\overline{\rho} \frac{D\tilde{k}}{Dt} = \vec{\nabla} \cdot \left[\overline{\rho} \left(\nu + \frac{\nu_t}{S_{ck}} \right) \vec{\nabla} \tilde{k} \right] - \overline{\rho} \widetilde{\vec{v}'' \vec{v}''} : \left(\vec{\nabla} \vec{\tilde{v}} \right) - \overline{\rho} \tilde{\epsilon}. \tag{4.110}$$

The main limitation of the Prandtl model is determining the scale of turbulent length, l, which becomes a significant problem for complex flows.

(b) Model of Two Equations $\tilde{k} - \tilde{\epsilon}$

The model $\tilde{k} - \tilde{\epsilon}$ is one of the most used in many commercial codes for turbulent flows. It considers the turbulent viscosity as a function of the turbulent kinetic energy $\tilde{k}$ and the turbulent dissipation $\tilde{\epsilon}$, according to equation

$$\nu_t = c_\mu \frac{\tilde{k}^2}{\tilde{\epsilon}}, \qquad c_\mu = 0.09. \tag{4.111}$$

From these two quantities, the length scale can be obtained from ($L = \tilde{k}^{3/2}/\tilde{\epsilon}$) and the time scale from ($\tau = \tilde{k}/\tilde{\epsilon}$). The model requires transport equations for $\tilde{k}$ and $\tilde{\epsilon}$, as given in equations:

$$\overline{\rho} \frac{D\tilde{k}}{Dt} = \vec{\nabla} \cdot \left[\overline{\rho} \left(\nu + \frac{\nu_t}{\sigma_k} \right) \vec{\nabla} \tilde{k} \right] - \overline{\rho} \widetilde{\vec{v}'' \vec{v}''} : \left(\vec{\nabla} \vec{\tilde{v}} \right) - \overline{\rho} \tilde{\epsilon}, \tag{4.112}$$

$$\overline{\rho} \frac{D\tilde{\epsilon}}{Dt} = \vec{\nabla} \cdot \left[\overline{\rho} \left(\nu + \frac{\nu_t}{\sigma_\epsilon} \right) \vec{\nabla} \tilde{\epsilon} \right] - C_{\xi_1} \overline{\rho} \frac{\tilde{\epsilon}}{\tilde{k}} \widetilde{\vec{v}'' \vec{v}''} : \left(\vec{\nabla} \vec{\tilde{v}} \right) - C_{\xi_2} \overline{\rho} \frac{\tilde{\epsilon}^2}{\tilde{k}}. \tag{4.113}$$

According to the standard model $\tilde{k} - \tilde{\epsilon}$, the parameters σ_k, σ_ϵ, C_{ξ_1}, and C_{ξ_2} have values 1, 1.3, 1.44, and 1.92, respectively [10].

In this formulation, the transport equations can be solved more easily numerically than those for the Reynolds tensor.

Equations (4.112) and (4.113) apply to free shear flows. For boundary layer flows, some modifications are necessary to take into account the presence of the wall. An alternative is to replace the boundary conditions at $y = 0$ per other, at a distance y_0 from the wall outside the viscous sublayer, in order to avoid integration of the equations in the region close to the wall surface, in which there are high gradients in y direction.

In general, y_0 is defined as

$$y_0 = \left(\frac{\nu}{u_\tau}\right) y_0^+, \tag{4.114}$$

where y_0^+ has a value of 50 for smooth surfaces, while u_τ is the friction velocity, given by

$$u_\tau = \left(\frac{\tau_W}{\rho}\right)^{1/2} \tag{4.115}$$

and τ_W is the shear stress over the wall.

To determine the velocity profile at y_0, the law of the wall can be used, which states that the average velocity of the turbulent flow at a point is proportional to the logarithm of the distance from that point from the wall. Thus, the expressions for the components of x and y of velocity vector in $y = y_0$ can be written according to

$$\tilde{u}_0 = u_\tau \left[\frac{1}{\kappa} \ln\left(\frac{y_0 u_\tau}{\nu}\right) + c\right], \tag{4.116}$$

$$\tilde{v}_0 = -\frac{u_0 y_0}{u_\tau} \frac{du_\tau}{dx}, \tag{4.117}$$

where c is a constant with values between 5 and 5.2 [10], while κ corresponds to the Von Kárman constant.

A well-known deficiency of the $\widetilde{k} - \widetilde{\epsilon}$ model is that this model overestimates the spreading rate of a circular jet. This problem can be remedied by modifying the values of the coefficients C_{ξ_1} and C_{ξ_2} [7]. Another way to avoid this problem is to use modified $\widetilde{k} - \widetilde{\epsilon}$ models, for example, the $\widetilde{k} - \tilde{w}$ model [13].

4.2.4 Models for Turbulent Viscosity in LES

Among the models for the turbulent viscosity commonly used in Large-Eddy Simulation (LES), we present those of Smagorinsky and Germano.

(a) Smagorinsky's Model

The model proposed by Smagorinsky [16] is one of the most used in LES. In this model, the eddy viscosity is assumed to be proportional to the sub-grid length scale (Δ) and the characteristic turbulent velocity ($V_\Delta = \Delta|\overline{S}|$) as

$$\nu_t = l^2|\overline{S}| = (C_s\Delta)^2\,|\overline{S}|, \tag{4.118}$$

where l is the length scale, C_s is the Smagorinsky coefficient, Δ is the filter size, and $\overline{S} = \sqrt{2\overline{S}_{i,j}\overline{S}_{i,j}}$ is the Frobenius norm of the strain rate.

In the Smagorinsky model, the turbulent effects follow approximately the cascade of energy of Kolmogorov, $k^{-5/3}$, in the sub-grid scale. This model is too dissipative, for quasi two-dimensional turbulent flows and near a wall [8].

(b) Germano's Model

The Germano's [17] model proposes the implementation of the Smagorinsky model for two filters:

- a mesh filter $\overline{\Delta}$,
- a test filter $\widehat{\Delta}$,

where the test filter is assumed to be greater than the mesh length scale. In this model, coefficient C_s varies in time and space.

Applying the filter in the equations results in the usual expressions of LES methodology, described in Chapter 6, with the residual tensor τ_{ij}^R given by the relation

$$\tau_{ij}^{\mathrm{R}} = \overline{\rho}\left(\overline{v_i v_j} - \overline{v_i}\,\overline{v_j}\right). \tag{4.119}$$

Filtering again the set of equations, but this time by applying the test filter, yields the following relation for the residual tensor T_{ij}^{R}:

$$T_{ij}^{\mathrm{R}} = \overline{\rho}\left(\widehat{\overline{v_i v_j}} - \widehat{\overline{v_i}}\,\widehat{\overline{v_j}}\right). \tag{4.120}$$

The residual stress tensor of the two filters relates to the identity of Germano as

$$\mathcal{L}_{ij} = T_{ij}^{\mathrm{R}} - \tau_{ij}^{\mathrm{R}}. \tag{4.121}$$

Substituting the expressions for the eddy viscosity given by the Smagorinsky model, the Germano identity (4.121) is obtained:

$$\mathcal{L}_{ij} = -2\overline{\rho}C_s M_{ij}, \qquad M_{ij} = (\hat{\Delta})^2\|\widehat{\overline{S_{ij}}}\|\widehat{\overline{S_{ij}}} - (\overline{\Delta})^2\widehat{\|\overline{S_{ij}}\|\overline{S_{ij}}}. \tag{4.122}$$

Equation (4.122) can be used to determine the value of C_s. However, because this is a tensor equation, the system of equations is overdetermined. Lilly [18] proposed an expression for C_s that best suits the system given by (4.122) by minimizing the least squares error, obtaining the equation

$$C_s^2 = \frac{1}{2}\frac{\mathcal{L}_{ij}M_{ij}}{M_{ij}^2}. \tag{4.123}$$

Germano's model is more efficient for describing the turbulence than Smagorinsky's model, but it implies in a high computational cost.

REFERENCES

[1] Peters N. Turbulent combustion. Cambridge, UK: Cambridge University Press; 2006.

[2] Peters N. Fifteen lectures on laminar and turbulent combustion; 1992, consulted in July 15, 2008 http://www.itv.rwth-aachen.de/fileadmin/LehreSeminar/Combustion/SummerSchool.pdf.

[3] Floyd J. Multi-parameter, multiple fuel mixture fraction combustion model for fire dynamics simulator. US Department of Commerce Building and Fire Research Laboratory, National Institute of Standards and Technology; 2008.

[4] Tennekes H, Lumley JL. A first course in turbulence. Cambridge, MA: MIT Press; 1972.

[5] Hinze JO. Turbulence: an introduction to its mechanism and theory. New York: McGraw-Hill; 1959.

[6] Poinsot T, Veynante D. Theoretical and numerical combustion. Philadelphia, PA: R.T. Edwards Inc.; 2001.

[7] Pope SB. Turbulent flows. Cambridge, UK: Cambridge University Press: 2000.

[8] Lesieur M. Turbulence in fluids. Berlin: Springer; 2008.

[9] Boussinesq J. Théorie de lécoulement tourbillant, Mem Priésentés par Divers Savants Acad Sci Inst 1877;23:46-50 [French].

[10] Cebeci T. Analysis of turbulent flows. Amsterdam: Elsevier; 2004.

[11] Agrawal A, Prasad A. Integral solution for the mean flow profiles of turbulent jets, plumes, and wakes, J Fluids Eng 2003;125:813-22.

[12] Schlichting H. Boundary-layer theory. New York: McGraw-Hill Book Company; 1979.

[13] Wilcox DC. Turbulence modeling for CFD. La Canada, CA: DCW Industries, Incorporated; 1994.

[14] Baldwin B, Barth TJ. A one-equation turbulence model for high Reynolds number wall-bounded flows. NASA Technical Memorandum 102847; 1990.

[15] Spalart PR, Allmaras SR. A one-equation turbulence model for aerodynamic flows. AIAA Paper 1992;92:0439.

[16] Smagorinsky J. General circulation experiments with the primitive equations. I. The basic experiment. Month Weather Rev 1963;91(3):99-164.

[17] Germano M, Piomelli U, Moin P, Cabot WH. A dynamic subgrid-scale eddy viscosity model. Phys Fluids 1991;3(7):1760-5.

[18] Lilly DK. A proposed modification of the Germano subgrid index-scale closure method. Phys Fluids 1992;4(3):633-5.

Chapter 5

Models for Reactive Flows

Reactive flows are present in a variety of applications, such as premixed and diffusion flames, flow in porous media with and without dissolution, and precipitation of minerals. These flows frequently include a large number of chemical reactions, increasing the computational work to analyze them.

5.1 SOME TECHNIQUES TO OBTAIN REDUCED MECHANISMS

Among many techniques used to obtain skeletons of larger mechanisms, in what follows we discuss the Direct Relation Graph (DRG) technique, coupled with Depth First Search (DFS) technique, the Sensitivity Analysis of the Jacobian matrix for the chemical system, the Intrinsic Low-Dimensional Manifold (ILDM) technique, the Reaction Diffusion Manifolds (REDIM technique) and the flamelet technique. Among these, the flamelet technique is preferred for writing the simplified chemical system for premixed flames and diffusion flames presented in the following sections.

5.1.1 Direct Relation Graph (DRG)

The numerical solution of flames and geochemical problems often involves hundreds of chemical species and a number of an order of magnitude greater than that of elementary reactions. Coupled with the presence of spray and formation of pollutants, the full numerical treatment of such models is still impractical. So, it is convenient to consider a reduced number of species and elementary reactions. This leads to the development of reduced kinetic mechanisms.

The techniques of reduction of chemical mechanisms can be classified in terms of global reduction in the number of reactions and of refinement of the mechanism. In this context, DRG is seen as a powerful technique to obtain skeleton mechanisms. If the production of species A depends on species B, it requires the presence of species B in the mechanism. This dependence is expressed by the index of importance

Modeling and Simulation of Reactive Flows. http://dx.doi.org/10.1016/B978-0-12-802974-9.00005-2

$$I_{AB} = \frac{\sum_{k=1}^{n} |\nu_{A_k} \dot{w}_k| \delta_{B_k}}{\sum_{k=1}^{n} |\nu_{A_k} \dot{w}_k|}, \tag{5.1}$$

where ν_{A_k} is the stoichiometric coefficient of species A in the reaction k, $\dot{w}_k$ is the difference between the forward and reverse reaction rates, and δ_{B_k} is a coefficient of value 1 if species B is present in the reaction k, or zero if species B does not belong to the reaction k. Thus, the value of I_{AB} ranges from zero to 1; when it approaches zero, this is an indication that the species is not very important, and if it does not belong to the main chain, this species can be eliminated from the mechanism. This technique is fast, easy to implement, and allows a skeletal mechanism to be obtained rapidly. For example, it only takes a few minutes to reduce a mechanism for biodiesel with 12,000 reactions to 600-1200 reactions [1]. A threshold value can be defined for I_{AB}, below which the species involved can be eliminated.

Variations of this technique, such as the Direct Relation Graph with Error Propagation (DRGEP) [2], allow for obtaining larger reductions at low cost. In particular, the DRG-aided Sensitivity Analysis (DRGASA) is associated with the sensitivity analysis or importance of the species involved [3]. Depending on the threshold parameter value, flaws in the main chain obtained by the DRG and related techniques may appear. Therefore, it is appropriate to use this technique with a search algorithm for graphs, as the Depth First Search (DFS), to establish the species of the main chain [4].

The main advantage of the use of DFS technique is that it allows, among the many possible routes to obtain a reduced mechanism, establishment of the main path to determine the principal combustion products (H_2O, CO_2, CO, and H_2). The species related to the main path can be used as target species when using the DRG techniques, reducing the computational effort to obtain the skeleton mechanism.

A representation of the DFS technique can be seen in Figure 5.1. The main steps for the DFS technique consist of:

1. Declare two search lists, one opened and other closed.
2. Add fuel, A, to the closed list and its sons, B and C, to the opened list.
   - Opened list: B, C
   - Closed list: A
3. Because B has a child, but is not one of the target species, add it to the closed list and its child, D, to the opened list. C has no children and is not one of the target species, so it remains in the opened list.
   - Opened list: C, D
   - Closed list: A, B
4. Because D has children, but is not one of the target species, add D to the closed list and its children, E and F, to the opened list.

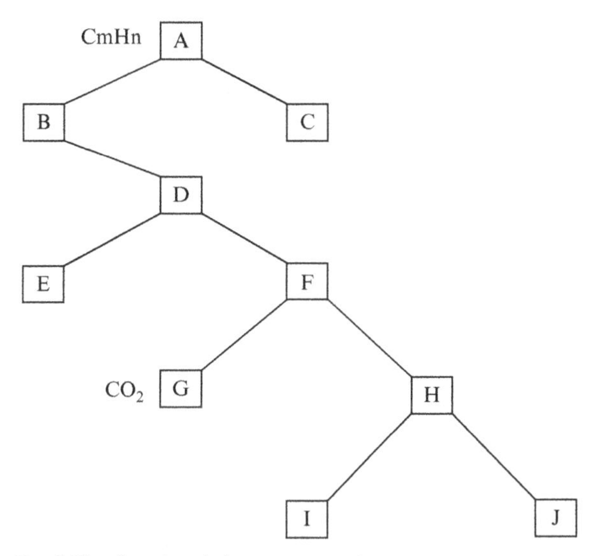

FIGURE 5.1 Depth First Search technique representation.

- Opened list: C, E, F
- Closed list: A, B, D

5. Because E has no children and is not one of the target species, it remains on the opened list. F has children, G and H, but is not one of the target species, so add F to the closed list and its sons, G and H, in the opened list.
   - Opened list: C, E, G, H
   - Closed list: A, B, D, F
6. Because G has no children, but is one of the target species, add it to the closed list. H has children, but it is not one of the target species, therefore add it to the closed list and the children to the opened list.
   - Opened list: C, E, I, J
   - Closed list: A, B, D, F, G, H
7. Because I and J have no children, and they do not correspond to the target species, both remain in the opened list. H does not have children and is not one of the target species, so H returns to the opened list.
   - Opened list: C, E, H, I, J
   - Closed list: A, B, D, F, G

Without further species to test, the species contained in the closed list correspond to the main chain.

This procedure varies according to the chemical scheme of interest. In addition, the main steps are written here a little differently than shown in the literature in order to facilitate their application.

When increasing the number of levels, the number of possible paths between the fuel and the products CO_2 and H_2O can be very large. The main chain is formed by paths whose rates are lower or whose reactions are dominant.

Because most of the computational work decreases when the number of differential equations is reduced, this reduction is proportional to the decrease in number of species. Each species is represented by an ordinary differential equation.

According to Pepiot and Pitsch (2005) [2], around 100 species are needed for the ignition of n-heptane. However, 12 species provide about 99% of mass of the final combustion products. Therefore, in order to reduce the computational cost, a skeletal mechanism with around 100 species is obtained and from these, reduced mechanisms are developed based on their reaction rates.

It is necessary to identify and maintain the species that are directly or indirectly coupled with the species of the main chain. However, if a step does not contribute to the rate of production and consumption of a species, this reaction may be removed from the mechanism.

The DRG method assumes that each selected species has equal importance and those species with I_{AB} greater than the cutoff value, and species related to them, are also selected. In addition, an other drawback of this technique is that the limit value (threshold) is not directly related to the error of the mechanism obtained.

5.1.2 Sensitivity Analysis

Because a detailed kinetic mechanism for conventional fuels consists of hundreds of species and thousands of elementary reactions, reduced mechanisms help to analyze three-dimensional reactive flows.

From the overall system, one can obtain a skeleton mechanism that includes only a minimal set of reactions and species that are necessary for the system description. From a skeleton mechanism, one can obtain a reduced mechanism, valid for the specific conditions for which it had been obtained. The sensitivity analysis method is one of the principal techniques used to obtain reduced kinetic mechanisms [3, 5, 6].

The sensitivity analysis is an important tool that evaluates the response of a model due to changes of one or more parameters. Among the sensitivity analysis methods are the Direct Sensitivity Analysis (DSA), Principal Component Analysis (PCA), Normalized Rate Sensitivity Coefficients (NRSC), and Overall Normalized Species Sensitivity Coefficients (ONSS) [7].

Following the DSA technique, consider the equation

$$\frac{d\vec{C}}{dt} = \vec{f}(\vec{C}, \vec{k}), \tag{5.2}$$

which corresponds to the temporal variation of species concentration. One obtains a system of differential equations for the local sensitivity coefficients from

$$\frac{\mathrm{d}}{\mathrm{d}t}\frac{\partial \vec{C}}{\partial k_j} = \frac{\partial \vec{f}}{\partial k_j} = J\frac{\partial \vec{C}}{\partial k_j} + \frac{\partial \vec{f}}{\partial k_j}, \tag{5.3}$$

where $J(t)$ is the Jacobian matrix, ($J = \partial\vec{f}/\partial\vec{C}$).

Because the parameters have different orders of magnitude, working with normalized coefficients yields the normalized sensitivity matrix

$$S_{ij} = \frac{\partial \ln(C_i)}{\partial \ln(k_j)}, \tag{5.4}$$

where C_i is the concentration of the species i and k_j the reaction rates of each j reaction. Reactions with sensitivity coefficients, S_{ij}, lower than the cut-off value can be excluded from the mechanism. Note that the natural logarithm can be regarded as a scale of normalization.

Equation (5.4) provides us the local sensitivities. A global sensitivity analysis can be calculated using the square of the sum of the normalized sensitivity matrix

$$B_j = \sum_{i}^{n} \left(\frac{\partial \ln(C_i)}{\partial \ln(k_j)} \right)^2. \tag{5.5}$$

Operations with the Jacobian matrix are known to consume high computational time in most simulations involving implicit solvers. Based on the sensitivity analysis of the Jacobian matrix, around two-third of reactions and one-third of chemical species can be eliminated from the complete mechanism without significant loss in the quality of results.

5.1.3 Intrinsic Low-Dimensional Manifold (ILDM) Technique

The ILDM technique seeks to identify those processes that have small time scales, which may be decoupled from the other processes [8]. This technique has some disadvantages, because it becomes more complex in regions of low temperature and transport (convection, diffusion, etc.). This technique relies on the decomposition of dynamic modes (eigenvalues) of low and high order. Because the high-order modes are rapidly damped, the processes are governed by low-order modes. Thus, a low number of modes is needed to describe the dynamics of the system, reducing the rigidity of the resulting system of reactive equations [9, 10].

The ILDM technique exploits the multi-scale structure of reactive flows to find a variety of low dimension modes that approximate the reactive processes in the thermochemical space. The reactive system of equations can be written as

$$\frac{\mathrm{d}\vec{\phi}}{\mathrm{d}t} = \vec{F}(\vec{\phi}), \tag{5.6}$$

where $\vec{\phi} = (T, p, Y_1, \ldots, Y_n)^{\mathrm{T}}$ is a vector composed of temperature, pressure, and mass fraction of species k. $\vec{F}(\vec{\phi})$ is the n-dimensional vector associated with the reactive source term.

The Jacobian matrix $J = (\partial\vec{F}/\partial\vec{\phi})$ is used to identify a decomposition of the states in subspaces of fast and slow reactions. Suppose the eigenvalues set is decomposed into m_s eigenvalues with small real part and $m_f = n - m_s$ eigenvalues with large real part, where n is the number of species.

Consider a subspace composed of points for which the reaction rates toward the m_f fast processes are canceled, resulting in the equation of the ILDM:

$$\vec{\tilde{Z}}_f(\vec{\phi})\vec{F}(\vec{\phi}) = 0. \tag{5.7}$$

This equation represents a system of m_f equations with n variables ($n > m_f$). So m_s equations need to be added to the system to ensure the uniqueness of the solution during the process of generating the ILDM. These equations are of the form

$$\vec{P}_z(\vec{\phi} - \vec{\phi}_P) = \vec{\alpha}_c, \tag{5.8}$$

$$\vec{P}_r(\vec{\phi} - \vec{\phi}_P) = \Delta\vec{\theta}, \tag{5.9}$$

where $\vec{P}_z$ is the n_z algebraic expressions used to set the conserved quantities (concentrations, enthalpy, and pressure) and $\vec{P}_r$ the $n_r = m_s - n_z$ expressions that parameterize the reactive processes. α is a vector n_z-dimensional characterizing the parameters of the conserved variables and $\vec{\phi}_P$ is a previous existing point in the manifold. Thus, the reduced mechanism represents the thermochemical state for a small number of variables, denoted by $\vec{\theta}$.

The solution procedure consists of:

1. Starting with an initial point, for example the point of equilibrium.
2. Generating a mesh m_s dimensional in the coordinate directions.
3. Solving Eqs. (5.8)–(5.9) in each node of the mesh.

To calculate $\vec{P}_r$, first is estimated

$$\vec{\phi}_{\theta_i} = \vec{\phi}_P(\vec{\theta}) - \vec{\phi}_P(\vec{\theta} - \vec{\delta}_i), \tag{5.10}$$

where $\vec{\delta}_i = (\delta_{i_1}, \ldots, \delta_{i_{n_r}})^T$ are the cell points, δ_{ij} the Kronecker delta, and $\vec{\phi}_\theta = (\phi_{\theta_1}, \ldots, \phi_{\theta_{nr}})^T$ is an approximation of the gradient ($n \times n_r$)-dimensional of $\vec{\phi}$ based on the cell vertices with respect to $\vec{\theta}$. The inclusion of $\vec{\phi}_\theta$ in the equation for $\vec{P}_r$ provides

$$\vec{P}_r \cdot \vec{\phi}_\theta = I, \tag{5.11}$$

where I is the identity matrix. Through the pseudo-inverse of Moore-Penrose $\vec{\phi}^n_\theta = [(\phi^T_\theta\vec{\phi}_\theta)^{-1}\vec{\phi}^T_\theta]$, $\vec{P}_r(\vec{\phi})$ may be defined as:

$$\vec{P}_r = \vec{\phi}^n_\theta. \tag{5.12}$$

The ILDM technique has been used to generate reduced mechanism for various combustion systems in the literature. Among its disadvantages are that:

- The ILDM does not always exist, or can be difficult to obtain, for example, in low-temperature regions.
- Transport is neglected in the generation of the ILDM.

5.1.4 Reaction Diffusion Manifolds (REDIM) Technique

Recently the REDIM technique, which combines the advantages of flamelet and ILDM, emerged. Because the REDIM technique directly couples the chemical and molecular transport, it can be seen as an extension to high order for the flamelet technique [11, 12].

When the thermochemical processes and the convective-reactive terms have the same importance, both must be included in the model to cover what occurs for reactive processes at low or moderate temperature.

The equations that model the chemistry and transport can be written in a generic way as

$$\frac{\partial \vec{\phi}}{\partial t} = \vec{f}(\vec{\phi}) - \vec{v} \cdot \vec{\nabla}\vec{\phi} - \frac{1}{\rho}\vec{\nabla} \cdot (D\vec{\nabla}\vec{\phi}) = \vec{F}(\vec{\phi}) \tag{5.13}$$

or

$$\frac{\partial \vec{\phi}}{\partial t} = \vec{F}(\vec{\phi}), \tag{5.14}$$

where $\vec{\phi} = (T, p, Y_1, \ldots, Y_n)^{\mathrm{T}}$ is a vector $n = 2 + n_{\mathrm{sp}}$-dimensional composed of the scalars temperature, pressure, and mass fraction of species k and ρ, which is the density. $\vec{f}(\vec{\phi})$ is the n-dimensional vector associated with the source term, $\vec{v}$ is the velocity vector, and D is the matrix of diffusion coefficients.

The dynamics of this system of equations is governed by time scales that may differ by several orders of magnitude. The scales that are very fast may be insignificant to model the numerical problem.

The concentration of some species evolves more rapidly than others for a certain initial period of time. When enough time passes, the terms of the slowest time scales begin to affect the system's behavior. Over time, the terms of slower time scales dominate the evolution of the species concentration.

When the slow and fast time scales of a reaction differ by several orders of magnitude, the most important information about system behavior is usually included in the terms of slow scales. Due to this, rapid changes in time scales can be assumed to be in steady state without significant loss in kinetic description of the system.

After a certain time, the dynamic of Eq. (5.13) has movements only in the m_{s} principal directions of space. The concept of invariant manifolds (U is invariant if $f(U) \subset U$) is applied in which the variety is defined by

$$\vec{M} = \left\{ \vec{\phi} : \vec{\phi} = \vec{\phi}(\vec{\theta}), \vec{\phi} : \mathbb{R}^{m_{\mathrm{s}}} \to \mathbb{R}^{n} \right\}, \tag{5.15}$$

where $\vec{\phi}(\vec{\theta})$ is an explicit function and $\vec{\theta}$ is a vector m_s dimensional of local coordinates, which parameterizes the variety.

The concept of invariant variety can be better understood considering that, for a $\vec{x}_0$ belonging to a set of points P in a n-dimensional space, the solution $\vec{x}(t)$ with initial condition $\vec{x}_0$ is in P for $t < T$, with $T > 0$. If this condition is valid for $T \rightarrow \infty$, then P is an invariant variety. So, for all $\vec{\phi} \in \vec{M}$, the vectorial field $\vec{F}(\vec{\phi}) \in \vec{T}_{\phi}\vec{M}$. In this way,

$$[\vec{\phi}_{\theta}^{n}(\vec{\theta})]^{\mathrm{T}} \cdot \vec{F}(\vec{\phi}) = 0 \tag{5.16}$$

for all $\vec{\theta}$ that defines $\vec{M}$, where $\vec{\phi}_{\theta}^{n}$ denotes the space normal to the variety, $(\vec{\phi}_{\theta}^{n})^{\mathrm{T}}\vec{\phi}_{\theta} = 0$.

This condition implies that

$$\vec{\vec{P}}^{n} \cdot \vec{F}(\vec{\phi}) = (\vec{\vec{I}} - \vec{\phi}_{\theta}\vec{\phi}_{\theta}^{n}) \cdot \vec{F}(\vec{\phi}) = 0, \tag{5.17}$$

where $\vec{\vec{P}}^{n} = (\vec{\vec{I}} - \vec{\phi}_{\theta}\vec{\phi}_{\theta}^{n})$ is an approximation of the operator projection of the normal space to $\vec{M}$. The pseudo inverse $\vec{\phi}_{\theta}^{n}$ is defined by

$$\vec{\phi}_{\theta}^{n} = \frac{\vec{\phi}_{\theta}^{T}}{\vec{\phi}_{\theta}^{T}\vec{\phi}_{\theta}}. \tag{5.18}$$

Based on the observation that in many reactive flows, the term $\vec{f}(\vec{\phi})$ is dominant, many procedures neglect the transport term in $\vec{F}(\vec{\phi})$, which is a good approximation for high temperatures.

Assuming that there is an explicit function $\vec{\phi}(\vec{\theta})$, $\vec{F}[\vec{\phi}(\vec{\theta})]$ can be written as

$$\vec{F}(\vec{\phi}(\vec{\theta})) = \vec{f}(\vec{\phi}(\vec{\theta})) - \vec{v} \cdot \vec{\phi}_{\theta}\vec{\nabla}(\vec{\theta}) - \frac{1}{\rho}\vec{\nabla} \cdot [D\vec{\phi}_{\theta}\vec{\nabla}(\vec{\theta})]. \tag{5.19}$$

Expanding the last term yields

$$\begin{aligned}\vec{F}(\vec{\phi}(\vec{\theta})) = {} & \vec{f}(\vec{\phi}(\vec{\theta})) - \vec{v} \cdot \vec{\phi}_{\theta}\vec{\nabla}(\vec{\theta}) - \frac{1}{\rho}[D\vec{\phi}_{\theta}(\vec{\nabla} \cdot \vec{\nabla}(\vec{\theta}))] \\ & -\frac{1}{\rho}[(D\vec{\phi}_{\theta})_{\theta} \cdot \vec{\nabla}(\vec{\theta}) \cdot \vec{\nabla}(\vec{\theta})].\end{aligned} \tag{5.20}$$

Equal diffusivities D are generally assumed for all species. Applying the operator $\vec{\vec{P}}^{n} = (\vec{\vec{I}} - \vec{\phi}_{\theta}\vec{\phi}_{\theta}^{n})$ in $\vec{F}[\vec{\phi}(\vec{\theta})]$ results in

$$(\vec{\vec{I}} - \vec{\phi}_{\theta}\vec{\phi}_{\theta}^{n})(\vec{v} \cdot \vec{\phi}_{\theta}\vec{\nabla}(\vec{\theta})) = \vec{v} \cdot \vec{\phi}_{\theta}\vec{\nabla}(\vec{\theta}) - \vec{v} \cdot (\vec{\phi}_{\theta}\vec{\phi}_{\theta}^{n}\vec{\phi}_{\theta})\vec{\nabla}(\vec{\theta}) = 0. \tag{5.21}$$

In this way, the projection over $\vec{F}[\vec{\phi}(\vec{\theta})]$ results in

$$(\vec{\vec{I}} - \vec{\phi}_{\theta}\vec{\phi}_{\theta}^{n}) \cdot \vec{F}(\vec{\phi}(\vec{\theta})) = (\vec{\vec{I}} - \vec{\phi}_{\theta}\vec{\phi}_{\theta}^{n}) \cdot \left(\vec{f}(\vec{\theta}) - \frac{d}{\rho}\vec{\phi}_{\theta\theta} \cdot \vec{\nabla}(\vec{\theta}) \cdot \vec{\nabla}(\vec{\theta})\right). \tag{5.22}$$

Therefore, using Eq. (5.17), one defines the reduced variety by

$$(\vec{\vec{I}} - \vec{\phi}_\theta \vec{\phi}_\theta^n) \cdot \left(\vec{f}(\vec{\theta}) - \frac{d}{\rho}\vec{\phi}_{\theta\theta} \cdot \vec{\nabla}(\vec{\theta}) \cdot \vec{\nabla}(\vec{\theta})\right) = 0, \tag{5.23}$$

and following Bykov and Maas [12], the problem is solved as

$$\frac{\partial(\vec{\phi}(\vec{\theta}))}{\partial t} = (\vec{\vec{I}} - \vec{\phi}_\theta \vec{\phi}_\theta^n) \cdot \left(\vec{f}(\vec{\theta}) - \frac{d}{\rho}\vec{\phi}_{\theta\theta} \cdot \vec{\nabla}(\vec{\theta}) \cdot \vec{\nabla}(\vec{\theta})\right) \tag{5.24}$$

with initial condition $\vec{\phi}_0$.

5.1.5 Flamelet Technique: REDIM in One Dimension

A simplified thermochemical process can be modeled by the equations

$$\frac{\partial T}{\partial t} = \frac{\dot{w}_\mathrm{T}}{\rho} - \vec{v} \cdot \vec{\nabla} T + \frac{1}{\rho}\vec{\nabla} \cdot (D\vec{\nabla} T), \tag{5.25}$$

$$\frac{\partial Y_i}{\partial t} = \frac{\dot{w}_i}{\rho} - \vec{v} \cdot \vec{\nabla} Y_i + \frac{1}{\rho}\vec{\nabla} \cdot (D\vec{\nabla} Y_i), \tag{5.26}$$

$$\frac{\partial p}{\partial t} = 0, \tag{5.27}$$

forming a $n = 2 + n_\mathrm{sp}$ dimensional system. This system can be written in vector form as

$$\frac{\partial \vec{\phi}}{\partial t} = \vec{f}(\vec{\phi}) - \vec{v} \cdot \vec{\nabla}\vec{\phi} + \frac{1}{\rho}\vec{\nabla} \cdot (D\vec{\nabla}(\vec{\phi})), \tag{5.28}$$

where $\vec{\phi} = (T, p, Y_\mathrm{F}, \ldots, Y_{n_\mathrm{sp}})^\mathrm{T}$ and $\vec{f}(\vec{\phi}) = \left(\frac{\dot{w}_\mathrm{T}}{\rho}, 0, \frac{\dot{w}_\mathrm{F}}{\rho}, \ldots, \frac{\dot{w}_\mathrm{sp}}{\rho}\right)^\mathrm{T}$. The diffusion matrix $D = \kappa/c_\mathrm{p}$ is a diagonal matrix, $\dot{w}_i$ is the rate of the reaction species i, $\dot{w}_\mathrm{T}$ is the rate of heat release due to the reaction, κ is the thermal conductivity, and c_p is the specific heat at constant pressure.

In the one-dimensional REDIM ($m_\mathrm{s} = 1$) it is assumed that the parametrization is made through only one variable. In what follows, consider $\vec{\theta}$ a vector of one element equal to Z, the mixture fraction that varies in the interval [0,1], based on the hypothesis of rapid chemistry. Therefore, the gradient vector, ϕ_θ is given by

$$\vec{\phi}_\theta = \left(\frac{\partial T}{\partial Z}, \frac{\partial p}{\partial Z}, \frac{\partial Y_\mathrm{F}}{\partial Z}, \ldots, \frac{\partial Y_{n_\mathrm{sp}}}{\partial Z}\right). \tag{5.29}$$

Assuming that $\vec{\phi}_\theta$ and $\vec{\phi}_{\theta\theta}$ transforms in

$$\vec{\phi}_\theta = \left(\frac{\partial T}{\partial Z}, 0, \frac{\nu_{\mathrm{st}} Y_{\mathrm{F}_i} + Y_{\mathrm{O}_{2_i}}}{\nu_{\mathrm{st}}}, \frac{\partial Y_2}{\partial Z}, \ldots, \frac{\partial Y_{n_{\mathrm{sp}}}}{\partial Z}\right) \tag{5.30}$$

and

$$\vec{\phi}_{\theta\theta} = \left(\frac{\partial^2 T}{\partial Z^2}, 0, 0, \frac{\partial^2 Y_2}{\partial Z^2}, \ldots, \frac{\partial^2 Y_{n_{\mathrm{sp}}}}{\partial Z^2}\right), \tag{5.31}$$

it results in $\vec{\phi}_\theta^n = \vec{\phi}_\theta^T / [\vec{\phi}_\theta^T \vec{\phi}_\theta]$, which is a n-dimensional matrix

$$\phi_\theta^n = \left(0, 0, \frac{\nu_{\mathrm{st}}}{\nu_{\mathrm{st}} Y_{\mathrm{F}_i} + Y_{\mathrm{O}_{2i}}}, 0, 0\right). \tag{5.32}$$

Thus, the REDIM system is given by the stationary solution of

$$\frac{\partial T}{\partial t} = \frac{\kappa}{\rho c_p} |\vec{\nabla} Z|^2 \frac{\partial^2 T}{\partial Z^2} + \frac{1}{\rho} \dot{w}_T, \tag{5.33}$$

$$\frac{\partial Y_k}{\partial t} = \frac{\kappa}{\rho c_p} |\vec{\nabla} Z|^2 \frac{\partial^2 Y_k}{\partial Z^2} + \frac{1}{\rho} \dot{w}_k, \tag{5.34}$$

or after introducing the scalar dissipation rate

$$\chi = \frac{2\lambda}{\rho c_{\mathrm{p}}} |\vec{\nabla} Z|^2, \tag{5.35}$$

provides

$$\frac{\partial T}{\partial t} = \frac{\rho \chi}{2} \frac{\partial^2 T}{\partial Z^2} + \frac{1}{\rho} \dot{w}_T, \tag{5.36}$$

$$\frac{\partial Y_k}{\partial t} = \frac{\rho \chi}{2} \frac{\partial^2 Y_k}{\partial Z^2} + \frac{1}{\rho} \dot{w}_k. \tag{5.37}$$

The transport is not neglected when the equation for the mixture fraction in Eulerian form is employed, namely

$$\frac{\partial Z}{\partial t} = -\vec{v}.\vec{\nabla} Z + \vec{\nabla}.\left(\frac{\kappa}{c_p} \vec{\nabla} Z\right) \tag{5.38}$$

To obtain the REDIM in more dimensions, species with high concentrations are frequently chosen, such as water vapor (H_2O), carbon dioxide (CO_2), and nitrogen (N_2).

5.2 MODELS FOR PREMIXED FLAMES

Premixed combustion requires that fuel and oxidant are mixed before they enter the combustion chamber. An example of a practical application is the spark ignition engine. The following sections discuss the characteristic velocity for laminar premixed flames, the governing equations, and the time and length scales for premixed flames.

5.2.1 Laminar Burning Velocity

In premixed combustion, the velocity at which the flame front propagates (normal to itself) is called laminar burning velocity, s_L. This is a thermochemical transport property that depends on the equivalence ratio between fuel and air, the temperature of the unburned mixture and the pressure.

In Figure 5.2 a Bunsen burner, which is a device for generating a laminar premixed flame, is shown. The gaseous fuel enters through an orifice in the mixing chamber, into which air enters via adjustable openings.

Figure 5.3 shows the kinematic balance to this process. The velocity vector of the unburned mixture v_u (subscribed u) is decomposed in $v_{t,\,u}$ component, which is tangential to the flame front, and $v_{n,\,u}$ component, which is normal to the flame front. Because there is a thermal expansion inside the flame front, the normal velocity component is increased, because the mass flow (ρv) due to the flame is the same in the unburned mixture and in the burned gas (subscript b), that is, $(\rho v_n)_u = (\rho v_n)_b$ and therefore

$$v_{n,b} = v_{n,u}\frac{\rho_u}{\rho_b}. \tag{5.39}$$

In Figure 5.3, the sum of the velocity of the components in the burned gas leads to the speed $\vec{v}_b$. Because the flame is in steady-state condition $s_{L,u} = v_{n,u}$ and due to angle α, represented in Figure 5.2, the normal speed becomes $v_{n,u} = v_u \cdot \sin\alpha$ and thus $s_{L,\,u} = v_u \sin\alpha$. This allows one to experimentally determine the burning rate by measuring the value of α, under the condition that the flow velocity v_u is uniform in the pipe outlet.

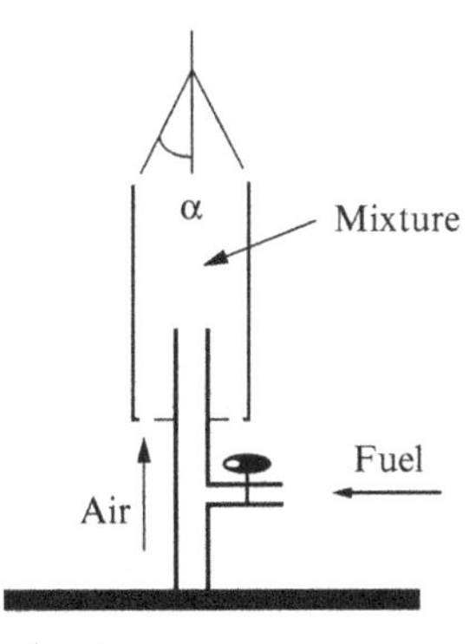

FIGURE 5.2 Sketch of a Bunsen burner.

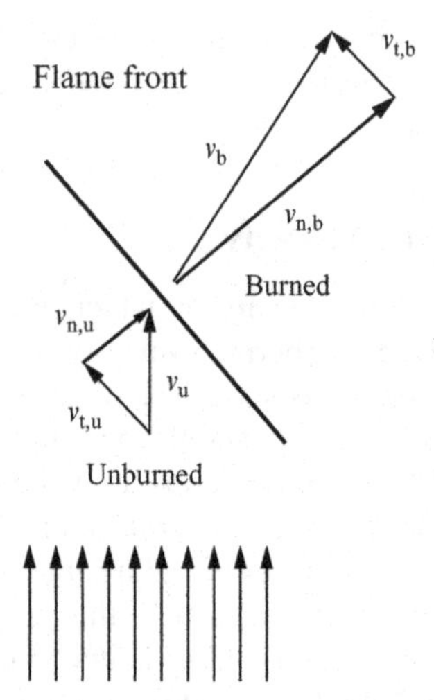

FIGURE 5.3 Kinetic balance for an oblique flame.

5.2.2 Governing Equations for Premixed Flames

Figure 5.4 shows the structure of a steady planar flame in the normal direction to the x axis with the unburned mixture in $x \to -\infty$ and the gas burned in $x \to +\infty$. Consider the case of one-step reaction

$$\nu'_F F + \nu'_{O_2} O_2 \to \nu''_P P. \tag{5.40}$$

The fuel and oxidant are transported with speed s_L with mass fractions $Y_{F,u}$ and $Y_{O_2,u}$ at $x \to -\infty$, and diffusion occurs within the reaction zone. In this case, the fuel is completely consumed, and the remaining oxygen is transferred back by convection. With the chemical reaction, product P is formed, releasing heat, and there is an increase of mass fraction Y_P. The heat conduction in the reaction zone causes a preheating of the fuel-air mixture.

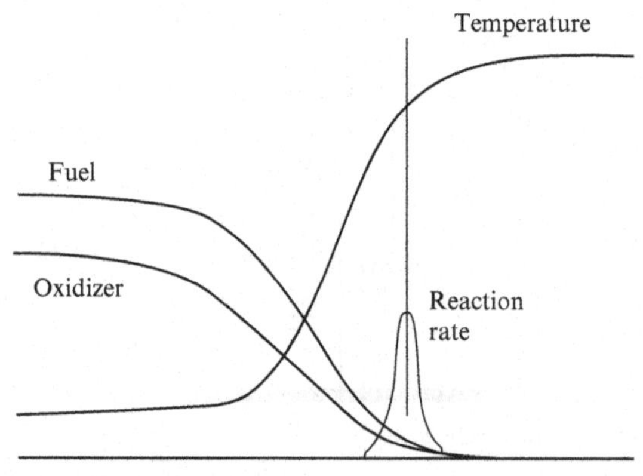

FIGURE 5.4 Structure of a premixed flame.

The burning velocity, s_L, can be calculated by solving the equations for total mass, species, and temperature, respectively,

$$\frac{d(\rho u)}{dx} = 0, \tag{5.41}$$

$$\rho u \frac{dY_i}{dx} = -\frac{dj_i}{dx} + \dot{w}_i, \tag{5.42}$$

$$\rho u c_p \frac{dT}{dx} = \frac{d}{dx}\left(\kappa \frac{dT}{dx}\right) - \sum_{i=1}^{n} h_i \dot{w}_i - \sum_{i=1}^{n} c_{p,i} j_i \frac{dT}{dx} + \frac{\partial p}{\partial t}. \tag{5.43}$$

Integrating the continuity equation (5.41) results in

$$\rho u = \rho_u s_L, \tag{5.44}$$

where the subscript u denotes the condition in the unburned mixture, and s_L denotes the burning velocity. For the propagation of the flame burning at a much lower speed than the speed of sound, the pressure is spatially almost constant and can be obtained by the equation of state. Only the spatial gradients of pressure are neglected in Eq. (5.43) and pressure gradients in time are kept.

Moreover, consider the description of the structure of a premixed flame developed by Zeldovich and Kamenetzki-Frank [13]. The reaction rate is of first order with respect to fuel and oxygen, and is given by

$$\dot{w} = B \frac{\rho Y_F}{W_F} \frac{\rho Y_{O_2}}{W_{O_2}} \exp\left(-\frac{E_a}{RT}\right). \tag{5.45}$$

This equation shows an Arrhenius type temperature dependence for high activation energy. Both the activation energy and the frequency factor B are parameters obtained experimentally.

Because the reaction is irreversible by hypothesis, the reaction rate cancels when the reagent is entirely consumed: the fuel for lean flame, the oxidant for rich flame, and both of them for stoichiometric flame. With this becomes the condition for the burned gas $Y_{F,b} \cdot Y_{O_2,b} = 0$. In the reaction zone there is an increase in temperature and, consequently, of the reaction rate. The diffusive flux is given by $\vec{j}_i = -\rho \mathcal{D}_{ij} \vec{\nabla} Y_i$ for the Lewis number equal to unity. For $\rho u = \rho_u s_L$, the equation of the mass fraction of species (5.42) of fuel and oxygen are [13]

$$\begin{aligned} \rho_u s_L \frac{dY_F}{dx} &= \frac{d}{dx}\left(\frac{\kappa}{c_p}\frac{dY_F}{dx}\right) - \nu'_F W_F \dot{w}, \\ \rho_u s_L \frac{dY_{O_2}}{dx} &= \frac{d}{dx}\left(\frac{\kappa}{c_p}\frac{dY_{O_2}}{dx}\right) - \nu'_{O_2} W_{O_2} \dot{w}, \end{aligned} \tag{5.46}$$

and the temperature equation becomes

$$\rho_u s_L \frac{dT}{dx} = \frac{d}{dx}\left(\frac{\kappa}{c_p}\frac{dT}{dx}\right) + \frac{Q}{c_p}\dot{w}. \tag{5.47}$$

Solving these equations results in

$$Y_F = -\frac{\nu'_F W_F c_p}{Q}(T - T_b) + Y_{F,b},$$
$$Y_{O_2} = -\frac{\nu'_{O_2} W_{O_2} c_p}{Q}(T - T_b) + Y_{O_2,b}, \tag{5.48}$$

where Q and c_p are assumed constant for simplicity. Because Eq. (5.45) involves Y_F, Y_{O_2}, and T, it follows that the reaction rate is a function of temperature.

5.2.3 Characteristic Length and Time Scales

The burning velocity can be approximated as [13]

$$s_L = \left(\frac{\kappa_b}{\rho_u c_p}\frac{1}{t_c}\right)^{1/2}, \tag{5.49}$$

where $\kappa/(\rho c_p)$ is the thermal diffusivity and t_c is the chemical time scale defined by

$$t_c = \frac{\rho_u c_P (T_b - T_u)^2 E^2}{\rho_b 2BR^2 T_b^4 A} \exp\left(\frac{E}{RT_b}\right). \tag{5.50}$$

The thickness of the flame can be approximated using the dimensionless scale $x^* = \rho_u s_L \int_0^x (c_p/\kappa)\mathrm{d}x$. If $x^* = 1$, then the flame thickness l_F becomes

$$l_F = \frac{(\frac{\kappa}{c_p})T_{ref}}{\rho_u s_L}. \tag{5.51}$$

Here, all relevant properties for the reaction rate are assessed in the temperature of the burned gas T_b. The time required for the flame to go across its thickness, called the flame time, is given by $t_F = l_F/s_L$. Comparing Eqs. (5.49)–(5.51), it is observed that the flame time t_F equals t_c for a kinetic of one step.

In the evaluation of the flame temperature properties, T_b is not realistic, and it is best to use the inner layer temperature (T^0). The stoichiometric values of l_F for methane-air flames, considering the atmospheric pressure, are of the order of 0.2 mm when $[(\kappa/c_p)_{T_{ref}}]/\rho_u \simeq 7 \times 10^{-5}\,\mathrm{m^2/s}$ and $s_L \simeq 0.4\,\mathrm{m/s}$.

Because κ/c_p is independent of pressure, for ideal gases, the density increases linearly with pressure. In hydrocarbon flames, the burning rate decreases approximately as p^{-m}, where the exponent m is around 0.5 for methane flames. Thus, the thickness of the flame varies with pressure as p^{m-1}.

5.2.4 *G*-Equation for Laminar Premixed Flame

The *G*-equation provides a geometrical description of the flame front [14]. In the *G*-equation model, the flame front is identified on surface G_0, separating burned and unburned gases. The G-equation model was developed for the corrugated flame and the thin reaction zones, what makes it attractive as a combustion model for the simulation of spark ignition engines.

The relationship $\mathrm{d}r_\mathrm{f}/\mathrm{d}t = v_\mathrm{u} + s_{\mathrm{L,u}}$ between the propagation velocity, the flow rate, and the burning rate can be generalized by introducing a vector $\vec{n}$ normal to the flame, and writing [13]

$$\left.\frac{\mathrm{d}\vec{x}}{\mathrm{d}t}\right|_\mathrm{f} = \vec{v} + s_\mathrm{L}\vec{n}, \tag{5.52}$$

where $\vec{x}_\mathrm{f}$ is the vector describing the flame position, $(\mathrm{d}\vec{x}/\mathrm{d}t)_\mathrm{f}$ is the flame propagation speed, $\vec{v}$ is the velocity vector of the mixture, and $\vec{n}$ the vector normal to the flame given by

$$\vec{n} = -\frac{\vec{\nabla} G}{|\vec{\nabla} G|}, \tag{5.53}$$

where $G(\vec{x}, t)$ is a scalar field

$$G(\mathbf{x}, t) = G_0. \tag{5.54}$$

G_0 represents the surface of the flame, as shown in Figure 5.5. $G > G_0$ corresponds to the burned gas region and $G < G_0$ to the unburned mixture. Differentiating Eq. (5.54) with respect to t in $G = G_0$ gives

$$\frac{\partial G}{\partial t} + \vec{\nabla} G \cdot \left.\frac{\partial \vec{x}}{\partial t}\right|_{G=G_0} = 0. \tag{5.55}$$

The G-equation is obtained by combining Eqs. (5.55) and (5.52),

$$\rho\left(\frac{\partial G}{\partial t} + \vec{v}\cdot\vec{\nabla} G\right) = \rho s_\mathrm{L}|\vec{\nabla} G|, \tag{5.56}$$

A solution of Eq. (5.56) is possible for variable values of s_L. Leaflets can be formed at different parts of the flame, and the flame eventually becomes fluctuating with time.

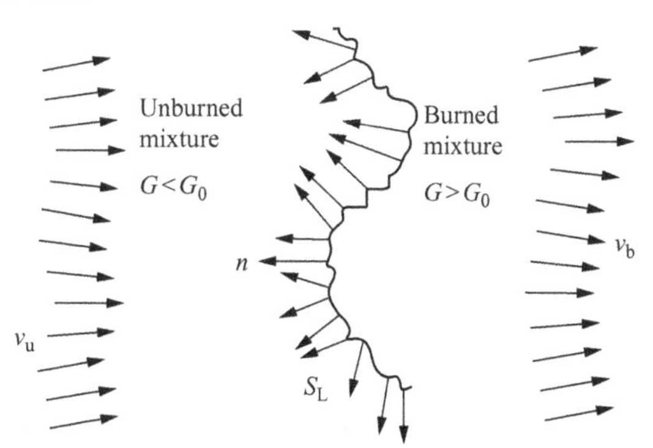

FIGURE 5.5 Scheme of a premixed flame propagating arbitrarily.

5.2.5 *G*-Equation for Premixed Turbulent Combustion

Consider the reference viscosity, ν_{ref}, composed of the product of the laminar burning velocity and the flame thickness,

$$\nu_{\text{ref}} = s_{\text{L}} l_{\text{F}}. \tag{5.57}$$

The turbulent Reynolds number is defined as

$$Re_{\text{t}} = \frac{v' l}{s_{\text{L}} l_{\text{F}}}, \tag{5.58}$$

the number of Damköhler as

$$Da_{\text{t}} = \frac{\tau}{t_{\text{F}}} = \frac{s_{\text{L}} l}{v' l_{\text{F}}}, \tag{5.59}$$

and the number of Karlovitz as

$$Ka = \frac{1}{Da_{\eta}} = \frac{t_{\text{F}}}{t_{\eta}} = \frac{{l_{\text{F}}}^2}{\eta^2} = \frac{{v_{\eta}}^2}{{s_{\text{L}}}^2}. \tag{5.60}$$

The latter of these numbers corresponds to the inverse of the number of Damköhler, defined with the Kolmogorov time scale. With these definitions, the following relations are given for v'/s_{L} and l/l_{F} in terms of the dimensionless numbers *Re*, *Da*, and *Ka*

$$\frac{v'}{s_{\text{L}}} = Re\left(\frac{l}{l_{\text{F}}}\right)^{-1} = Da^{-1}\left(\frac{l}{l_{\text{F}}}\right) = Ka^{2/3}\left(\frac{l}{l_{\text{F}}}\right)^{1/3}. \tag{5.61}$$

Figure 5.6 shows the diagram of Borghi for premixed combustion in terms of the logarithm of v'/s_{L} versus the logarithm of l/l_{F} [15]. The lines $Re_{\text{t}} = 1$, $Da = 1$, and $Ka = 1$ represent boundaries between the different regimes of premixed turbulent combustion.

Among the four regimes, the flamelet regime is characterized by the inequalities $Re_{\text{t}} > 1$ (turbulence), $Da > 1$ (chemical fast), and $Ka < 1$ (shear sufficiently low).

In the regime characterized by $Re_{\text{t}} > 1$, $Da > 1$, and $Ka > 1$, the last inequality indicates that the smaller vortices can enter the flame structure ($\eta < l_{\text{F}}$). The broken reaction zone, at the top left side of the diagram, is characterized by $Re_{\text{t}} > 1$, $Ka > 1$, but $Da < 1$, indicating that the chemical reaction is slow compared to the turbulence.

The flamelet regime is divided into two regimes. The first regime is $v' < s_{\text{L}}$, where v' can be interpreted as the rotation velocity of large vortices. These vortices cannot penetrate the flame front. In the second regime, when $Ka < 1$ results $v' \geq s_{\text{L}} \geq v_{\eta}$. Because the speed of large vortices is greater than the burning rate, these vortices can distort the flame front.

The local *G*-equation is given by [14]

$$\frac{\partial G}{\partial t} + \vec{v} \cdot \vec{\nabla} G = s_{\text{L}} |\vec{\nabla} G|. \tag{5.62}$$

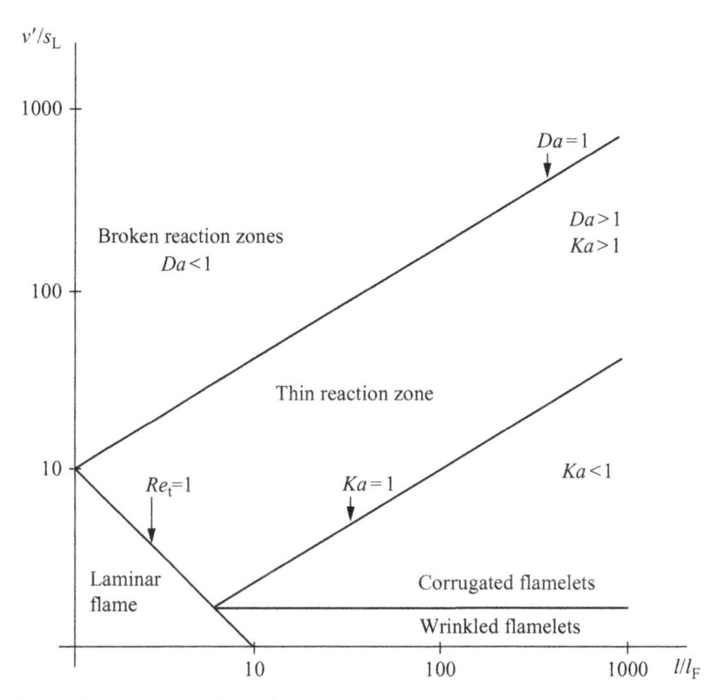

FIGURE 5.6 Phase diagram of premixed turbulent combustion.

Using Favre averages, by splitting G in

$$G = \tilde{G} + G'' \tag{5.63}$$

and using the mass conservation yields the G-equation for turbulent combustion [16]

$$\bar{\rho}\frac{\partial \tilde{G}}{\partial t} + \bar{\rho}\vec{v} \cdot \vec{\nabla}\tilde{G} = \bar{\rho}s_{\mathrm{T}}|\vec{\nabla}\tilde{G}| - \bar{\rho}D_{\mathrm{T}}k|\vec{\nabla}\tilde{G}|, \tag{5.64}$$

where the flame curvature is given by

$$k = \vec{\nabla} \cdot \vec{n} \tag{5.65}$$

with

$$\vec{n} = -\frac{\vec{\nabla}\tilde{G}}{|\vec{\nabla}\tilde{G}|}, \tag{5.66}$$

and s_{T} is the turbulent burning velocity, which may be approximated as

$$s_{\mathrm{T}} = s_{\mathrm{L}}(1 + \sigma_{\mathrm{T}}), \tag{5.67}$$

where σ_{T} is the increase of flame surface due to turbulence effects. The G-equation for premixed turbulent flames may be written in other forms too [17].

Because the pre-mixture of gases is frequently not complete, the study of diffusion flames is the next step.

5.3 MODELS FOR DIFFUSION FLAMES

In this section we describe some models for solving diffusion flames as the *flamelet* technique, a model for an axisymmetric jet diffusion flame and a model for a plume. Figure 5.7 shows the structure of a diffusion flame.

5.3.1 Flamelet Equations for Diffusion Flames

The flamelet concept for turbulent combustion applies when the reaction is fast compared to the mixture at the molecular level. In this regime, the chemistry of a flame and the turbulence can be treated separately. The flamelet concept approaches the solution of Burke-Schumann for a high Damköhler number and mechanism of one step. The scalar dissipation rate, which appears in the flamelet equations, relates the effects caused by the diffusion and convection. This rate is large at the smallest scales, but its fluctuations are mainly governed by the large scales, which are solved using Large-Eddy Simulation (LES).

When the flamelet model is applied to a diffusion flame, generally it is assumed that the molecular diffusivities are all equal ($D_Z = D_i = D_T$) and the numbers of Lewis are equal to unity ($Le_i = D_T/D_i = \kappa/(\rho c_p D_i) = 1$), thus introducing a common diffusivity defined as

$$D = \frac{\kappa}{\rho c_p}, \tag{5.68}$$

where κ is the thermal conductivity. These assumptions are more beneficial to hydrogen flames [13]. Although for hydrogen the Lewis number is low (~0.3), the flame temperature is high (~2300 k).

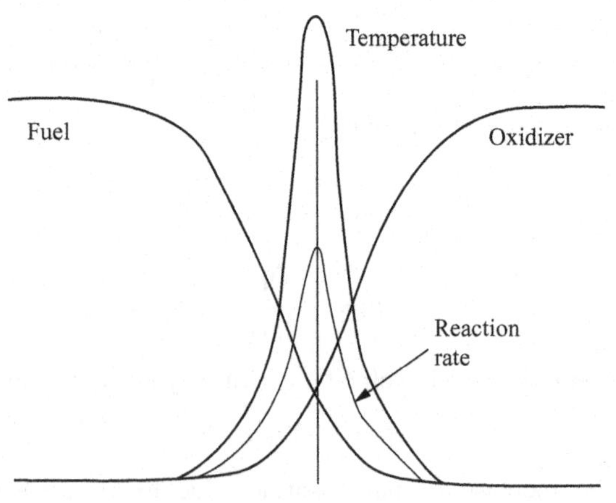

FIGURE 5.7 Structure of a diffusion flame.

With these considerations, the equations for the mixture fraction, mass fraction of species, and temperature, have the forms

$$\frac{\partial(\rho Z)}{\partial t}+\frac{\partial(\rho v_j Z)}{\partial x_j}=\frac{\partial}{\partial x_j}\left(\rho D\frac{\partial Z}{\partial x_j}\right), \tag{5.69}$$

$$\frac{\partial(\rho Y_i)}{\partial t}+\frac{\partial(\rho v_j Y_i)}{\partial x_j}=\frac{\partial}{\partial x_j}\left(\rho D\frac{\partial Y_i}{\partial x_j}\right)\pm\dot{w}_i, \tag{5.70}$$

$$\frac{\partial(\rho T)}{\partial t}+\frac{\partial(\rho v_j T)}{\partial x_j}=\frac{\partial}{\partial x_j}\left(\rho D\frac{\partial T}{\partial x_j}\right)+\dot{w}_\mathrm{T}. \tag{5.71}$$

The chemical source term and the rate of heat generation are given [15], respectively, by

$$\dot{w}_i = W_i\sum_{k=1}^{r}\nu_{ik}\dot{w}_k, \tag{5.72}$$

$$\dot{w}_\mathrm{T}=\frac{1}{c_\mathrm{P}}\sum_{k=1}^{r}Q_k\dot{w}_k, \tag{5.73}$$

where W_i is the molecular weight of species i, r the number of chemical reactions of the mechanism, ν_{ik} the stoichiometric coefficient of the species in the reaction k, c_P the specific heat at constant pressure, and $Q_k=-\sum_{i=1}^{n}\nu_{ik}W_ih_i$ the heat of combustion of the reaction k. The rate of the reaction is given by the equation

$$\dot{w}_j = k_{fj}\prod_{i=1}^{n}\left(\frac{\rho Y_i}{W_i}\right)^{\nu_{fij}}-k_{bj}\prod_{i=1}^{n}\left(\frac{\rho Y_i}{W_i}\right)^{\nu_{bij}}. \tag{5.74}$$

The flamelet equations describe the reactive-diffusive structure in the vicinity of the flame surface as a function of the mixture fraction [15]. Initially, a coordinate transformation is applied to the surface of the flame, then an analysis of the magnitude of terms is used to show that the derivatives of the reactive scalar in the tangential direction can be neglected when compared with the derivative in the normal direction.

For the transformation of coordinates, it is assumed that the surface of the flame is defined as the surface of a stoichiometric mixture $Z(x_j,t)=Z_\mathrm{st}$. If the local gradient of mixture fraction is sufficiently high, the combustion takes place in a thin layer in the vicinity of the stoichiometric surface. Figure 5.8 shows that the coordinate x_1, defined as being locally normal to the surface of the flame, is replaced by Z, while the tangential coordinates x_2 and x_3 are made equal to Z_2 and Z_3, respectively, and t is replaced by τ. Thus, Z is locally normal to the surface of stoichiometric mixture [13].

So, after applying the transformations

$$\frac{\partial}{\partial t}=\frac{\partial}{\partial \tau}+\frac{\partial Z}{\partial t}\frac{\partial}{\partial Z}, \tag{5.75}$$

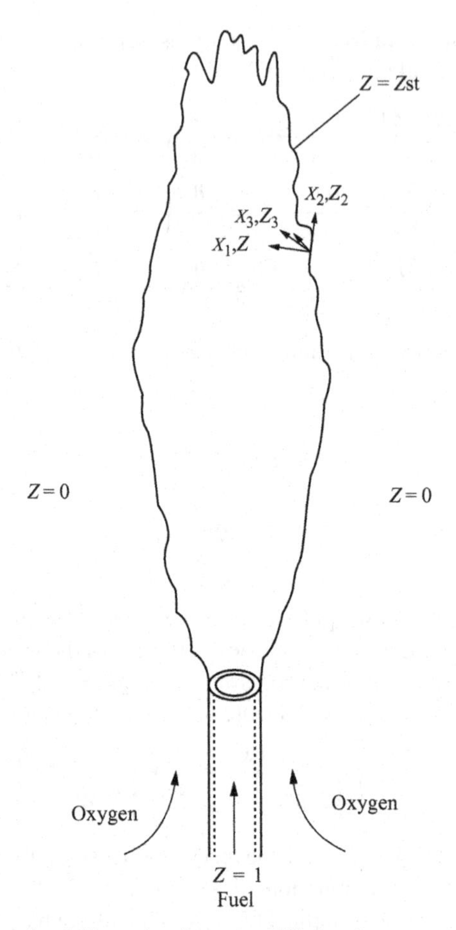

FIGURE 5.8 Surface of stoichiometric mixture in a diffusion flame.

$$\frac{\partial}{\partial x_k} = \frac{\partial}{\partial Z_k} + \frac{\partial Z}{\partial x_k}\frac{\partial}{\partial Z}, \quad k = 2, 3, \tag{5.76}$$

$$\frac{\partial}{\partial x_1} = \frac{\partial Z}{\partial x_1}\frac{\partial}{\partial Z}. \tag{5.77}$$

Equations (5.70) and (5.71) become

$$\begin{aligned}
&\frac{\partial(\rho Y_i)}{\partial \tau} + \frac{\partial Z}{\partial t}\frac{\partial(\rho Y_i)}{\partial Z} + \frac{\partial Z}{\partial x_1}\frac{\partial(\rho v_1 Y_i)}{\partial Z} + \frac{\partial(\rho v_k Y_i)}{\partial Z_k} + \frac{\partial Z}{\partial x_k}\frac{\partial(\rho v_k Y_i)}{\partial Z} \\
&= \frac{\partial Z}{\partial x_1}\frac{\partial}{\partial Z}\left(\rho D\left(\frac{\partial Z}{\partial x_1}\frac{\partial Y_i}{\partial Z}\right)\right) + \frac{\partial}{\partial Z_k}\left(\rho D\left(\frac{\partial Y_i}{\partial Z_k} + \frac{\partial Z}{\partial x_k}\frac{\partial Y_i}{\partial Z}\right)\right) \\
&+ \frac{\partial Z}{\partial x_k}\frac{\partial}{\partial Z}\left(\rho D\left(\frac{\partial Y_i}{\partial Z_k} + \frac{\partial Z}{\partial x_k}\frac{\partial Y_i}{\partial Z}\right)\right) \pm \dot{w}_i
\end{aligned} \tag{5.78}$$

and

$$\frac{\partial(\rho T)}{\partial \tau} + \frac{\partial Z}{\partial t}\frac{\partial(\rho T)}{\partial Z} + \frac{\partial Z}{\partial x_1}\frac{\partial(\rho v_1 T)}{\partial Z} + \frac{\partial(\rho v_k T)}{\partial Z_k} + \frac{\partial Z}{\partial x_k}\frac{\partial(\rho v_k T)}{\partial Z}$$
$$= \frac{\partial Z}{\partial x_1}\frac{\partial}{\partial Z}\left(\rho D\left(\frac{\partial Z}{\partial x_1}\frac{\partial T}{\partial Z}\right)\right) + \frac{\partial}{\partial Z_k}\left(\rho D\left(\frac{\partial T}{\partial Z_k} + \frac{\partial Z}{\partial x_k}\frac{\partial T}{\partial Z}\right)\right) \tag{5.79}$$
$$+ \frac{\partial Z}{\partial x_k}\frac{\partial}{\partial Z}\left(\rho D\left(\frac{\partial T}{\partial Z_k} + \frac{\partial Z}{\partial x_k}\frac{\partial T}{\partial Z}\right)\right) + \dot{w}_{\mathrm{T}}.$$

Because the density variation in the flow equations is more important than in the chemical equations, ρ can be considered constant in the chemical equations. For constant D, and using the equation of mixture fraction (5.69), one sees the following results:

$$\rho\left(\frac{\partial Y_i}{\partial \tau} + v_k\frac{\partial Y_i}{\partial Z_k}\right) = \frac{\partial}{\partial Z_k}\left(\rho D\frac{\partial Y_i}{\partial x_k}\right)$$
$$+ \rho D\left[\left(\frac{\partial Z}{\partial x_j}\right)^2\frac{\partial^2 Y_i}{\partial Z^2} + 2\frac{\partial Z}{\partial x_k}\frac{\partial^2 Y_i}{\partial Z\partial Z_k} + \frac{\partial^2 Y_i}{\partial Z_k^2}\right] \pm \dot{w}_i, \tag{5.80}$$

$$\rho\left(\frac{\partial T}{\partial \tau} + v_k\frac{\partial T}{\partial Z_k}\right) = \frac{\partial}{\partial Z_k}\left(\rho D\frac{\partial T}{\partial x_k}\right)$$
$$+ \rho D\left[\left(\frac{\partial Z}{\partial x_j}\right)^2\frac{\partial^2 T}{\partial Z^2} + 2\frac{\partial Z}{\partial x_k}\frac{\partial^2 T}{\partial Z\partial Z_k} + \frac{\partial^2 T}{\partial Z_k^2}\right] + \dot{w}_{\mathrm{T}}. \tag{5.81}$$

The hypothesis of the boundary layer is used in the analysis of the magnitude of the terms of these equations. The mass fraction of species and temperature are virtually constant along the stoichiometric surface. Gradients along the surface should be small, compared to the gradients in normal direction. Thus, neglecting the derivatives in the x_k and Z_k directions yields

$$\rho\frac{\partial Y_i}{\partial \tau} = \rho D\left(\frac{\partial Z}{\partial x_j}\right)^2\frac{\partial^2 Y_i}{\partial Z^2} \pm \dot{w}_i, \tag{5.82}$$

$$\rho\frac{\partial T}{\partial \tau} = \rho D\left(\frac{\partial Z}{\partial x_j}\right)^2\frac{\partial^2 T}{\partial Z^2} + \dot{w}_{\mathrm{T}}. \tag{5.83}$$

By defining the scalar dissipation rate as

$$\chi = 2D\left(\frac{\partial Z}{\partial x_j}\right)^2, \tag{5.84}$$

the following flamelet equations for mass fraction of species and temperature are produced:

$$\rho \frac{\partial Y_i}{\partial \tau} = \rho \frac{\chi}{2} \frac{\partial^2 Y_i}{\partial Z^2} \pm \dot{w}_i, \tag{5.85}$$

$$\rho \frac{\partial T}{\partial \tau} = \rho \frac{\chi}{2} \frac{\partial^2 T}{\partial Z^2} + \dot{w}_{\mathrm{T}} \tag{5.86}$$

with $\dot{w}_i$ and $\dot{w}_{\mathrm{T}}$ defined in Eqs. (5.72) and (5.73), respectively.

These equations complete the Lagrangian flamelet model. A transformation of coordinates different from that presented in Eqs. (5.75)–(5.77) results in the Eulerian flamelet model proposed by Pitsch [18]. In the Eulerian system, both velocity vector and scalar dissipation rate are functions of time, space, and the mixture fraction. The difference between these models appears to be the manner in which the fluctuations are taken into account. Because the differences are small, the Lagrangian flamelet model is more employed, because it is easier to implement and represents well the majority applications for diffusion flames.

Presumed Probability Density Function (pdf)

A probability density function serves to represent a probability distribution in terms of integrals [15]. Probability density functions, introduced in the Reynolds Averaged Navier-Stokes (RANS) context, are easily extended to Large-Eddy Simulation (LES), both for species mass fractions as well as for reaction rates. There are many alternatives to select a general class of shapes for the probability density functions of conserved scalars; however, a small number of parameters is desired [14].

The most typical pdf is a Gaussian function. The probability density function of the normal distribution with mean μ and variance σ^2 (standard deviation σ) is a Gaussian function:

$$f(x; \mu, \sigma) = \frac{1}{\sigma\sqrt{2\pi}} \mathrm{e}^{-(x-\mu)^2/2\sigma^2} = \frac{1}{\sigma}\phi\left(\frac{x-\mu}{\sigma}\right) \tag{5.87}$$

with the density function $\phi(x) = \frac{1}{\sqrt{2\pi}}\mathrm{e}^{-x^2/2}$.

The variance is a measure of the statistical dispersion, indicating how the possible values are spread around the expected value. The variance of a random variable X is $\sigma^2 = E(X-\mu)^2$ with the expected value $E(X) = \int_\Omega X\,\mathrm{d}P$.

The β-pdf is widely accepted as a good alternative function:

$$\tilde{P}(Z; x, t) = Z^{\alpha-1}(1-Z)^{\beta-1}\frac{\Gamma(\alpha+\beta)}{\Gamma(\alpha)\Gamma(\beta)} \tag{5.88}$$

for $0 \leq Z \leq 1$, where the gamma function is defined as $\Gamma(x) = \int_0^\infty \mathrm{e}^{-t} t^{x-1}\mathrm{d}t$, and $\alpha = \tilde{Z}\gamma$, $\beta = (1-\tilde{Z})\gamma$ and $\gamma = \frac{\tilde{Z}(1-\tilde{Z})}{\widetilde{Z''^2}} - 1 \geq 0$. This pdf can be evaluated using an extension of the Stirling's formula [19],

$$\Gamma(x) = (2\pi)^{1/2} x^{x-1/2} e^{-x+1/(12x)}. \tag{5.89}$$

The β functions are preferred to presume pdf because they are able to change continuously from pdf shapes to delta shapes [20]. For $\alpha < 1$ a singularity is developed at $Z = 0$; for $\beta < 1$ a singularity is developed at $Z = 1$. For such shapes, also found in jets, a composite model of three parts (a fully turbulent, a sublayer, and an outer flow) was developed by Effelsberg and Peters [19].

Because the pdf provides the statistical information about the variables, it has the ability to treat finite-rate chemistry and the turbulence-chemistry interactions.

The Scalar Dissipation Rate

The coupling between chemical kinetics and turbulence occurs through the scalar dissipation rate, which can be approximated using RANS or LES as follows, respectively:

$$\chi \sim C_\chi \frac{\tilde{\epsilon}}{\tilde{k}} \widetilde{Z''^2} \sim C_s^2 \frac{|\tilde{S}|}{S_c} \widetilde{Z''^2}. \tag{5.90}$$

Z''^2 may be approximated as

$$\tilde{Z}''^2 \sim \tilde{Z}(1 - \tilde{Z}) \tag{5.91}$$

or obtained by solving a differential equation [13].

The scalar dissipation rate acts as an external parameter that is imposed on the flamelet structure by the mixture fraction field [15]. It describes the influence of the turbulent flow field on the laminar flame structure. Both the mixture fraction and the scalar dissipation rate fluctuate on turbulent flows, and their statistical distribution needs to be considered. If the joint pdf $\tilde{P}(Z, \chi_{st})$ (where χ_{st} is χ at the stoichiometric condition) is known, the Favre mean of $\tilde{Y}_i$ can be obtained from

$$\tilde{Y}_i(x,t) = \int_0^1 \int_0^\infty Y_i(Z, \chi_{st}) \tilde{P}(Z, \chi_{st}; x, t)\, d\chi_{st}\, dZ. \tag{5.92}$$

The local scalar dissipation rate can be seen as a characteristic local diffusion time scale imposed by the mixing field [21].

For the counterflow diffusion flame (essentially a one-dimensional diffusion flame structure, as shown in Figure 5.9), one can write [15] $\chi(Z) = \frac{a}{\pi}(j+1)\exp\{-2[erfc^{-1}(2Z)]^2\} \sim 4aZ_{st}{}^2[erfc^{-1}(2Z_{st})]^2$, where $erfc(x) = 1 - erf(x)$ and $erf(x) = \frac{2}{\sqrt{\pi}} \int_0^x e^{-t}\, dx$. For a non-constant density, it yields the scalar dissipation rate the relation $\chi(Z) = \frac{a_\infty}{4\pi} \frac{3}{2} \left(\frac{\sqrt{\rho_\infty/(\rho+1)}}{2\sqrt{\rho_\infty/(\rho+1)}} \right) \exp\{-2[erfc^{-1}(2Z)]^2\}$ [22].

Here $a = -\partial v_\infty / \partial y$ is the potential velocity gradient in the oxidizer stream, $j = 0$ for planar flow, and $j = 1$ for axially symmetric flow. Because $\chi(Z)$ is the

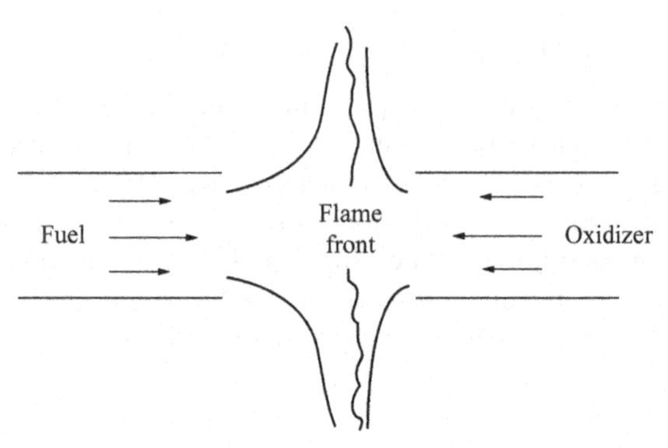

FIGURE 5.9 Counterflow diffusion flame.

same for various laminar flows and its exact value does not matter outside the reaction zone, the idea is to employ such expressions for other flows.

Heat Capacity

The heat capacity is a quantity that characterizes the ability of a body to store heat as it changes its temperature. It is common to employ NASA polynomials [23] for the temperature dependence on c_P for many different species. These polynomials have the following general form:

$$c_P/R = a_1 + a_2T + a_3T^2 + a_4T^3 + a_5T^4 \tag{5.93}$$

and the coefficients are given by the temperature usually on the following ranges: $300 < T < 1000$ and $T > 1000$.

For most hydrocarbon-air flames, the properties of nitrogen dominate, and the heat capacity of the mixture is very close to that of nitrogen, whose value changes from 1000 to 1300 J/kg K for temperatures between 300 and 3000 K [20].

5.3.2 The Burke-Schumann Solution

The Burke-Schumann solution for laminar diffusion flames uses the hypothesis of infinitely fast chemistry, valid for a high Damköhler number. With this hypothesis, the reaction occurs in a thin layer in the vicinity of the stoichiometric surface, separating the flame in to rich and lean portions. With this assumption, it is convenient to express the mass fraction of the components according to the mixture fraction. The mixture fraction can be defined as:

$$Z = \frac{Y_{F,u}}{Y_{F,1}}, \quad (1 - Z) = \frac{Y_{O_2,u}}{Y_{O_2,2}}, \tag{5.94}$$

where $Y_{F,u}$ and $Y_{O_2,u}$ correspond to the mass fractions of the fuel and the oxidizer in the unburned mixture, while $Y_{F,1}$ and $Y_{O_2,2}$ correspond the mass fractions of the fuel and oxidant in free streams, respectively.

Consider the one-step reaction given by

$$\upsilon_F F + \upsilon_{O_2} O_2 \rightarrow \upsilon_P P, \tag{5.95}$$

where F is the fuel and υ_F and υ_{O_2} the stoichiometric coefficients of fuel and oxygen, respectively. The relationship between the variations of the fuel and oxidant mass fractions can be written as

$$\frac{dY_F}{\upsilon_F W_F} = \frac{dY_{O_2}}{\upsilon_{O_2} W_{O_2}}, \tag{5.96}$$

where W_i is the molecular weight of component i.

Integrating Eq. (5.96) from unburned mixture ($t = 0$) to time t, one obtains the equation

$$\upsilon Y_F - Y_{O_2} = \upsilon Y_{F,u} - Y_{O_2,u}, \quad \upsilon = \frac{\upsilon_{O_2} W_{O_2}}{\upsilon_F W_F}. \tag{5.97}$$

Substituting the relationship given by Eq. (5.94) in Eq. (5.97) yields the mixture fraction in terms of mass fractions of fuel and oxidizer:

$$Z = \frac{\upsilon Y_F - Y_{O_2} + Y_{O_2,2}}{\upsilon Y_{F,1} + Y_{O_2,2}}. \tag{5.98}$$

For the case in which the fuel and oxidant are mixed in the stoichiometric level, $Y_F = \upsilon Y_{O_2}$, simplifies Eq. (5.98) according to:

$$Z_{st} = \left(1 + \upsilon \frac{Y_{F,1}}{Y_{O_2,2}}\right)^{-1}, \tag{5.99}$$

where Z_{st} corresponds to the stoichiometric mixture fraction.

For a lean flame, ($Z < Z_{st}$), combustion ends when all the fuel is consumed ($Y_{F,b} = 0$). The mass fraction of oxidant can be calculated according to (see Figure 5.10)

$$Y_{O_2,b} = Y_{O_2,2}\left(1 - \frac{Z}{Z_{st}}\right), \quad Z < Z_{st}. \tag{5.100}$$

For a rich flame, ($Z > Z_{st}$), the reaction ends when all the oxygen is consumed ($Y_{O_2,b} = 0$). Thus, the mass fraction of fuel can be determined as

$$Y_{F,b} = Y_{F,1}\left(\frac{Z - Z_{st}}{1 - Z_{st}}\right), \quad Z > Z_{st}. \tag{5.101}$$

For a hydrocarbon fuel, the one-step reaction can be written as

$$C_mH_n + \left(m + \frac{n}{4}\right) O_2 \rightarrow \frac{n}{2} H_2O + m\, CO_2. \tag{5.102}$$

Then, we can write the concentrations of H_2O and CO_2 as a function of the mixture fraction according to

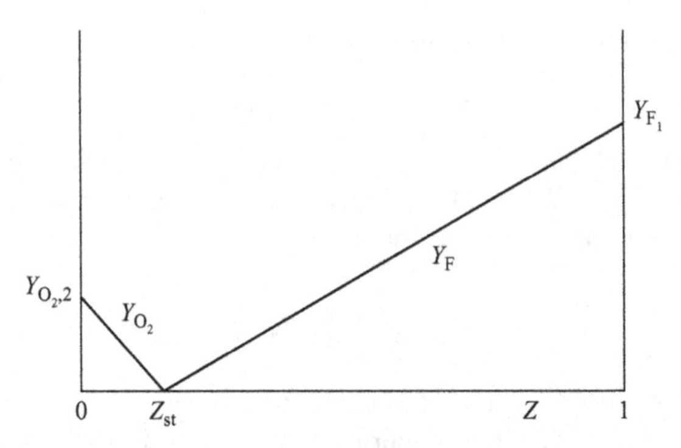

FIGURE 5.10 Burke-Schumann solution for Y_F and Y_{O_2} in the mixture fraction space.

$$Y_{H_2O,b} = \begin{cases} Y_{H_2O,st}\left(\dfrac{Z}{Z_{st}}\right), & Z \leq Z_{st} \\ Y_{H_2O,st}\left(\dfrac{1-Z}{1-Z_{st}}\right), & Z > Z_{st} \end{cases} \tag{5.103}$$

$$Y_{CO_2,b} = \begin{cases} Y_{CO_2,st}\left(\dfrac{Z}{Z_{st}}\right), & Z \leq Z_{st} \\ Y_{CO_2,st}\left(\dfrac{1-Z}{1-Z_{st}}\right), & Z > Z_{st} \end{cases} \tag{5.104}$$

with Y_{H_2O} and Y_{CO_2} stoichiometric defined by:

$$Y_{H_2O,st} = Y_{F,1}\, Z_{st}\, \frac{nW_{H_2O}}{2W_F}, \tag{5.105}$$

$$Y_{CO_2,st} = Y_{F,1}\, Z_{st}\, \frac{mW_{CO_2}}{W_F}. \tag{5.106}$$

The profiles of mass fractions of H_2O and CO_2 given by the Burke-Schumann solution are shown in Figure 5.11.

Similarly, it is possible to obtain an analytical solution for the temperature as a function of the mixture fraction. For an adiabatic system ($dq = 0$) at constant pressure ($dp = 0$) and with zero friction work ($dw_R = 0$), we have $dh = 0$. Thus, integrating the enthalpy of the unburned mixture to the final state of combustion yields

$$h_u = h_b, \tag{5.107}$$

leading to

$$\sum_{i=1}^{n} Y_{i,u} h_{i,u} = \sum_{i=1}^{n} Y_{i,b} h_{i,b}. \tag{5.108}$$

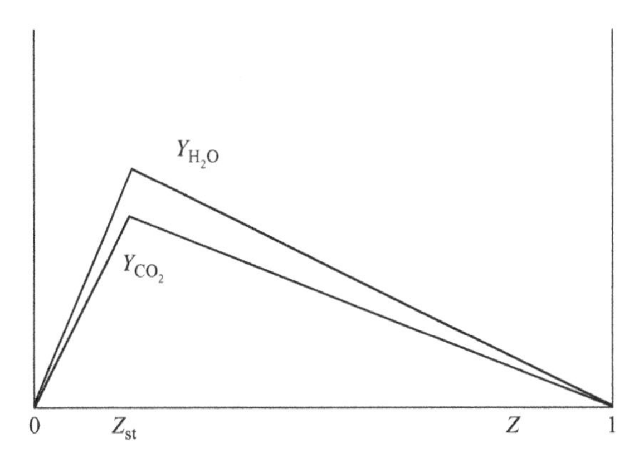

FIGURE 5.11 Burke-Schumann solution for Y_{H_2O} and Y_{CO_2} in the mixture fraction space.

Considering the dependence of the enthalpy on temperature given by Eq. (5.109), Eq. (5.108) can be rewritten according to [13]

$$h_i = h_{i,\mathrm{ref}} + \int_{T_{\mathrm{ref}}}^{T} c_{\mathrm{p}i}\, \mathrm{d}T, \tag{5.109}$$

$$\sum_{i=1}^{n} \left(Y_{i,\mathrm{u}} - Y_{i,\mathrm{b}}\right) h_{i,\mathrm{ref}} = \int_{T_{\mathrm{ref}}}^{T_{\mathrm{b}}} c_{\mathrm{p,b}}\, \mathrm{d}T - \int_{T_{\mathrm{ref}}}^{T_{\mathrm{u}}} c_{\mathrm{p,u}}\, \mathrm{d}T. \tag{5.110}$$

The specific heat capacities $c_{\mathrm{p,b}}$ and $c_{\mathrm{p,u}}$ can be obtained from the mass fraction of the components in the mixture, according to

$$c_{\mathrm{p,b}} = \sum_{i=1}^{n} Y_{i,\mathrm{b}}\, c_{\mathrm{p}_i}; \quad c_{\mathrm{p,u}} = \sum_{i=1}^{n} Y_{i,\mathrm{u}}\, c_{\mathrm{p}_i}. \tag{5.111}$$

Considering the global reaction given by Eq. (5.95), the variation of the mass fraction of a compound i due to the variation of the fuel mass fraction is given by

$$Y_{i,\mathrm{u}} - Y_{i,\mathrm{b}} = \left(Y_{\mathrm{F,u}} - Y_{\mathrm{F,b}}\right) \frac{\nu_i W_i}{\nu_{\mathrm{F}} W_{\mathrm{F}}}. \tag{5.112}$$

After multiplying Eq. (5.112) by $h_{i,\mathrm{ref}}$ and applying the sum for all components of the reaction, the result is

$$\sum_{i=1}^{n} \left(Y_{i,\mathrm{u}} - Y_{i,\mathrm{b}}\right) h_{i,\mathrm{ref}} = \frac{\left(Y_{\mathrm{F,u}} - Y_{\mathrm{F,b}}\right)}{\nu_{\mathrm{F}} W_{\mathrm{F}}} \sum_{i=1}^{n} \nu_i W_i h_{i,\mathrm{ref}}. \tag{5.113}$$

The heat of combustion can be defined according to

$$Q = -\sum_{i=1}^{n} \nu_i W_i h_i = -\sum_{i=1}^{n} \nu_i H_i. \tag{5.114}$$

To simplify the formulation, the heat of combustion is considered constant with the temperature and equal to Q_{ref}. $c_{p,b}$ is also considered to be constant and $T_u = T_{ref}$.

$$Q_{ref} = -\sum_{i=1}^{n} \nu_i H_{i,ref}. \tag{5.115}$$

Considering a lean flame ($Y_{F,b} = 0$) and combining Eqs. (5.110), (5.113), and (5.115), an expression is obtained for the temperature as a function of the mass fraction of the unburned fuel:

$$T_b - T_u = \frac{Q_{ref} Y_{F,u}}{c_p \nu_F W_F}. \tag{5.116}$$

For a rich flame ($Y_{O_2,b} = 0$), it becomes necessary to evaluate the variation of the compound i in relation to the variation of the oxygen mass fraction given by

$$Y_{i,u} - Y_{i,b} = \left(Y_{O_2,u} - Y_{O_2,b}\right) \frac{\nu_i W_i}{\nu_{O_2} W_{O_2}}. \tag{5.117}$$

Using the same procedure employed in Eq. (5.116), we obtain an expression for the temperature in the rich portion of the flame as

$$T_b - T_u = \frac{Q_{ref} Y_{O_2,u}}{c_p \nu_{O_2} W_{O_2}}. \tag{5.118}$$

Substituting the mixture fraction definitions given by Eq. (5.94) in Eqs. (5.116) and (5.118), the Burke-Schumann solution for the temperature can be written as:

$$T_b = \begin{cases} T_u + \dfrac{Q\, Y_{F,1}}{c_p \nu_F W_F} Z, & Z \leq Z_{st}, \\ \\ T_u + \dfrac{Q\, Y_{O_2,2}}{c_p \nu_{O_2} W_{O_2}} (1 - Z), & Z > Z_{st}, \end{cases} \tag{5.119}$$

$$T_u = T_2 + (T_1 - T_2)\, Z, \tag{5.120}$$

where T_1 is the temperature of the fuel stream, while T_2 is the temperature of the oxidant. Figure 5.12 shows the Burke-Schumann solution for temperature for a diffusion flame.

5.3.3 An Analytical Solution for Jet Flames

A jet corresponds to a free shear flow, in which a stream of high fluid velocity is forced under pressure through an orifice or nozzle. Jets can be classified according to the geometry gave them origin. A jet is rounded when generated by a circular orifice, or planar when generated by a rectangular cavity.

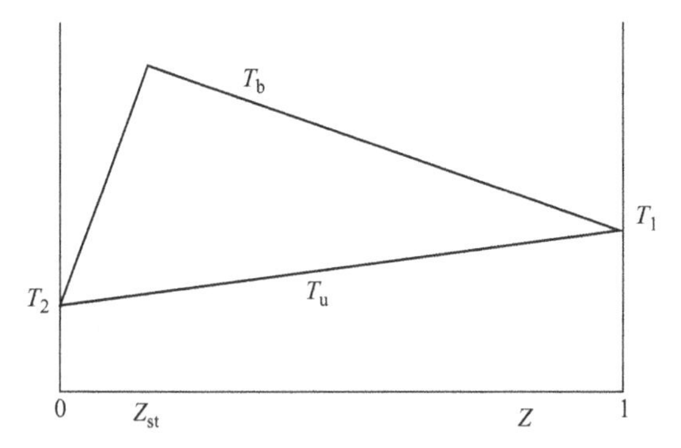

FIGURE 5.12 Burke-Schumann solution for temperature in the mixture fraction space.

Assuming a homogeneous system where a stream of fuel is mixed with an oxidant, the mixture fraction can be defined as the ratio between the fuel mass fraction in the unburned mixture and the fuel mass fraction in the original fuel stream.

$$\tilde{Z} = \frac{\tilde{Y}_{F,u}}{Y_{F,1}}. \tag{5.121}$$

For this analysis, consider a diffusion flame in which the fuel that exits from a nozzle with diameter d and speed u_0, is mixed with ambient air by convection and diffusion to form a jet flame, as shown in Figure 5.13.

The geometry of a jet is representative of the class of diffusion flames. This structure leads to a two-dimensional axisymmetric problem, governed by the equations of continuity, momentum, and mixture fraction in terms of Favre averages. These equations can be written as

$$\frac{\partial \overline{\rho}}{\partial t} + \frac{\partial \left(\overline{\rho}\tilde{v}_j\right)}{\partial x_j} = 0, \tag{5.122}$$

$$\overline{\rho}\left(\frac{\partial \tilde{v}_i}{\partial t} + \tilde{v}_j \frac{\partial \tilde{v}_i}{\partial x_j}\right) = -\frac{\partial \overline{p}}{\partial x_i} + \frac{\partial \overline{\tau_{ij}}}{\partial x_j} - \frac{\partial \left(\overline{\rho}\widetilde{v_j'' v_i''}\right)}{\partial x_j} + \overline{\rho} g_i, \tag{5.123}$$

$$\overline{\rho}\frac{\partial \tilde{Z}}{\partial t} + \overline{\rho}\tilde{v}_j \frac{\partial \tilde{Z}}{\partial x_j} = \overline{\frac{\partial}{\partial x_j}\left(\rho \frac{\nu}{Sc}\frac{\partial Z}{\partial x_j}\right)} - \frac{\partial \left(\overline{\rho}\widetilde{v_j'' Z''}\right)}{\partial x_j}, \tag{5.124}$$

where Sc is the Schmidt number.

To simplify these equations, the following hypotheses are adopted:

1. Steady state hypothesis, in which time derivatives are zero.
2. The effects of buoyancy and pressure gradient were neglected;
3. Boundary layer approximation, in which the second-order derivatives in the axial direction were neglected.

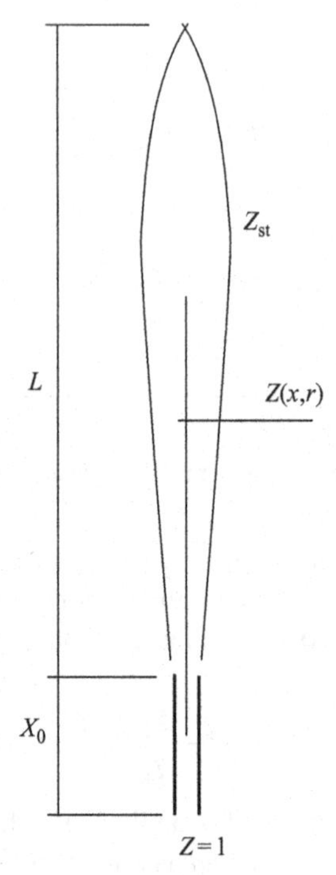

FIGURE 5.13 Schematic representation of a diffusive flame.

4. The viscous stress component $(\overline{\tau_{ij}})$ was neglected when compared to Reynolds component $\left(\overline{\rho}\widetilde{v_j''v_i''}\right)$.

5. The molecular component of the mixture fraction $\overline{\frac{\partial}{\partial x_j}\left(\rho \frac{\nu}{Sc}\frac{\partial Z}{\partial x_j}\right)}$ was neglected when compared to the turbulent component $\frac{\partial\left(\overline{\rho}\widetilde{v_j Z''}\right)}{\partial x_j}$.

To complete the system of equations, Boussinesq's hypothesis is assumed, in which the Reynolds stress acts in a similar way to the viscous stress, being proportional to the velocity gradient multiplied by a proportionality coefficient [24], $-\overline{\rho}\widetilde{u''v''} = \overline{\rho}\nu_t\frac{d\tilde{u}}{\partial y}$. This proportionality coefficient corresponds to the turbulent viscosity ν_t.

Similarly, the turbulent component of the mixture fraction, $\overline{\rho}\widetilde{v_j Z''}$, is modeled as being proportional to the gradient of $\tilde{Z}$, according to

$$-\overline{\rho}\widetilde{v''Z''} = \frac{\overline{\rho}\nu_t}{Sc} r \frac{\partial \tilde{Z}}{\partial r}. \tag{5.125}$$

With this simplification, the following equations in cylindrical coordinates [13] are produced:

$$\frac{\partial(\overline{\rho}\tilde{u}r)}{\partial x} + \frac{\partial(\overline{\rho}\tilde{v}r)}{\partial r} = 0, \tag{5.126}$$

$$\overline{\rho}\tilde{u}r\frac{\partial \tilde{u}}{\partial x} + \overline{\rho}\tilde{v}r\frac{\partial \tilde{u}}{\partial r} = \frac{\partial}{\partial r}\left(\overline{\rho}\nu_{\mathrm{T}}r\frac{\partial \tilde{u}}{\partial r}\right), \tag{5.127}$$

$$\overline{\rho}\tilde{u}r\frac{\partial \tilde{Z}}{\partial x} + \overline{\rho}\tilde{v}r\frac{\partial \tilde{Z}}{\partial r} = \frac{\partial}{\partial r}\left(\frac{\overline{\rho}\nu_{\mathrm{T}}r}{Sc}\frac{\partial \tilde{Z}}{\partial r}\right), \tag{5.128}$$

subject to the boundary conditions

$$\begin{aligned} r = 0: &\quad \tilde{v} = 0, \quad \tilde{Z} = \tilde{Z}_{\mathrm{cl}}, \\ r \to \infty: &\quad \tilde{u} = 0, \quad \frac{\partial \tilde{u}}{\partial r} = 0, \quad \tilde{Z} = 0. \end{aligned} \tag{5.129}$$

Under these conditions, for $r = 0$, the radial velocity is zero and the mixture fraction is equal to $\tilde{Z}_{\mathrm{cl}}$, while in $r \to \infty$, the axial velocity and its first order derivative are zero and the mixture fraction is zero due to lack of fuel.

The continuity equation (5.126) is satisfied by introducing the stream function

$$\overline{\rho}\tilde{u}r = \frac{\partial \psi}{\partial r}, \quad \overline{\rho}\tilde{v}r = -\frac{\partial \psi}{\partial x}. \tag{5.130}$$

The system of equations is simplified by applying the similarity transformation given by [25]

$$\eta = \frac{\overline{r}}{\zeta}, \quad \overline{r}^2 = 2\int_0^r \frac{\overline{\rho}}{\rho_\infty} r\,\mathrm{d}r, \quad \zeta = x + x_0. \tag{5.131}$$

The transformation given in Eq. (5.131) gives rise to a weighted radial coordinate, $\overline{r}$. The new axial coordinate ζ corresponds to the sum of the axial coordinate x and the virtual origin of the jet, x_0.

With this similarity transformation, the coordinate system (x, r) is transformed to (ζ, η) by the chain rule

$$\frac{\partial}{\partial x} = \frac{\partial}{\partial \zeta} + \frac{\partial \eta}{\partial \zeta}\frac{\partial}{\partial \eta}, \quad \frac{\partial}{\partial r} = \frac{\partial \eta}{\partial \overline{r}}\frac{\partial \overline{r}}{\partial r}\frac{\partial}{\partial \eta}. \tag{5.132}$$

The dimensionless stream function is given by

$$F = \frac{\psi}{\rho_\infty \nu_{\mathrm{t}} \zeta}. \tag{5.133}$$

The axial and radial velocities can be written as a function of F as

$$\tilde{u} = \frac{1}{\eta\zeta}\frac{\partial(\nu_{\mathrm{t}}F)}{\partial \eta}, \quad \overline{\rho}\tilde{v}r = -\rho_\infty\left[\nu_{\mathrm{t}}F - \eta\frac{\partial(\nu_{\mathrm{t}}F)}{\partial \eta}\right]. \tag{5.134}$$

The variable ν_t, present in Eqs. (5.133) and (5.134), corresponds to the turbulent viscosity of a jet with constant density. To simplify the calculations, it is assumed that the mixture fraction corresponds to the product of two functions: the mixture fraction in the jet centerline $\tilde{Z}_{cl}$, depending only on ζ, and the variable ω

$$\tilde{Z} = \tilde{Z}_{cl}\ \omega. \tag{5.135}$$

Substituting Eqs. (5.134) and (5.135) in Eqs. (5.127) and (5.128) and applying the chain rule, given by Eq. (5.132) yields the following equations [13]:

$$-\frac{\partial}{\partial \eta}\left[\frac{(\nu_t F)}{\eta}\frac{\partial(\nu_t F)}{\partial \eta}\right] = \frac{\partial}{\partial \eta}\left[C\nu_t\eta\frac{\partial}{\partial \eta}\left(\frac{1}{\eta}\frac{\partial(\nu_t F)}{\partial \eta}\right)\right], \tag{5.136}$$

$$-\frac{\partial}{\partial \eta}\left[(\nu_t F)\,\omega\right] = \frac{\partial}{\partial \eta}\left(\frac{C}{Sc}\eta\nu_t\frac{\partial \omega}{\partial \eta}\right), \tag{5.137}$$

subject to the following boundary conditions, obtained from the conditions expressed in Eq. (5.129),

$$\begin{aligned}
\eta = 0: &\quad F = 0, \quad (\nu_t F)' = 0, \quad \omega = 1, \\
\eta \to \infty: &\quad (\nu_t F)' = 0, \quad (\nu_t F)'' = 0, \quad \omega = 0.
\end{aligned} \tag{5.138}$$

The parameter C that appears in Eqs. (5.136) and (5.137) corresponds to the Chapman-Rubesin parameter described according to the equation

$$C = \frac{\overline{\rho}^2 \nu_T r^2}{\rho_\infty^2 \nu_t \overline{r}^2}. \tag{5.139}$$

According to Peters and Donnerhack [25], the Chapman-Rubesin parameter can also be written as

$$C = \frac{(\rho_0 \rho_{st})^{1/2}}{\rho_\infty}. \tag{5.140}$$

To quantify the eddy viscosity, one uses the semi-empirical expression obtained by Agrawal and Prasad [26] for an axisymmetric jet diffusion flame given by

$$\nu_t = \frac{U_c c^2}{4} x \left[\frac{1 - \exp\left(-\xi^2\right)}{\xi^2}\right], \quad \xi = \frac{r}{cx}, \tag{5.141}$$

where the variable U_c corresponds to the axial velocity on the centerline of the jet, which for an axisymmetric jet varies proportionally with x^{-1} [27].

To solve the differential equations (5.136) and (5.137) analytically using Agrawal and Prasad's [26] expression, the exponential term of Eq. (5.141) is approximated by a Taylor series. The expansion of the exponential term of Eq. (5.141) in a Taylor series truncated at the second term gives

$$\exp\left(-\xi^2\right) \approx 1 - \xi^2. \tag{5.142}$$

With this consideration, the eddy viscosity can be written according to equation

$$\nu_t = \frac{U_c c^2}{4} x. \tag{5.143}$$

Thus, the eddy viscosity becomes independent of η, because U_c is a function only of ζ. Thus, Eqs. (5.136) and (5.137) can be simplified, resulting in the equations

$$-\frac{\partial}{\partial \eta}\left(\frac{F}{\eta}\frac{\partial F}{\partial \eta}\right) = \frac{\partial}{\partial \eta}\left[C\eta \frac{\partial}{\partial \eta}\left(\frac{1}{\eta}\frac{\partial F}{\partial \eta}\right)\right], \tag{5.144}$$

$$-\frac{\partial}{\partial \eta}(F\omega) = \frac{\partial}{\partial \eta}\left(\frac{C}{Sc}\eta \frac{\partial \omega}{\partial \eta}\right). \tag{5.145}$$

Solving analytically Eq. (5.144) results in the following expression for the dimensionless stream function:

$$F = C(\gamma\eta)^2 \left[1 + \frac{(\gamma\eta)^2}{4}\right]^{-1}, \tag{5.146}$$

where parameter γ is constant with respect to η, obtained by integrating Eq. (5.144). The substitution of Eq. (5.146) in (5.134) leads to the following expression for the axial velocity in terms of ζ and η:

$$\tilde{u} = \frac{2\nu_t C\gamma^2}{\zeta}\left[1 + \frac{(\gamma\eta)^2}{4}\right]^{-2}. \tag{5.147}$$

This expression indicates that the speed at the center of the jet is proportional to x^{-1}, which is consistent with the initial consideration made for U_c in Eq. (5.141).

Combining the equation of momentum toward x Eq. (5.127) with the continuity equation (5.126) and integrating it from $r = 0$ to $r \to \infty$, gives

$$\frac{\partial}{\partial x}\left[\int_0^\infty \overline{\rho}\tilde{u}^2 r\,\mathrm{d}r\right] + \left[\overline{\rho}\tilde{u}\tilde{v}r\right]_0^\infty = \left[\overline{\rho}\nu_t r \frac{\partial \tilde{u}}{\partial r}\right]_0^\infty. \tag{5.148}$$

Analyzing the boundary conditions given by Eq. (5.129), it can be concluded that the last two terms in Eq. (5.148) are zero and, therefore, the integral within the first term is independent of x. Considering a top-hat profile for the velocity in $x = 0$ yields the equation

$$\int_0^\infty \overline{\rho}\tilde{u}^2 r\,\mathrm{d}r = \frac{\rho_0 u_0^2 R^2}{2}, \tag{5.149}$$

where ρ_0 is the density of the fuel stream at the nozzle exit, and R is the orifice radius.

Substituting Eq. (5.147) in Eq. (5.149) and integrating, the value of γ is determined, whose expression corresponds to

$$\gamma^2 = \frac{3}{64}\left(\frac{\rho_0}{\rho_\infty C^2}\right)\left(\frac{u_0 d}{\nu_t}\right)^2. \tag{5.150}$$

Substituting Eq. (5.146) into Eq. (5.145), and integrating Eq. (5.145) results in the expression for ω

$$\omega = \left[1 + \frac{(\gamma\eta)^2}{4}\right]^{-2Sc}. \tag{5.151}$$

Combining the equation of mixture fraction (5.128) with the continuity equation (5.126), and integrating from $r = 0$ to $r \to \infty$ yields

$$\frac{\partial}{\partial x}\left[\int_0^\infty \overline{\rho}\tilde{u}\tilde{Z} r \, \mathrm{d}r\right] + \left[\overline{\rho}\tilde{v}\tilde{Z}r\right]_0^\infty = \left[\frac{\overline{\rho}\nu_t r}{Sc}\frac{\partial \tilde{Z}}{\partial r}\right]_0^\infty. \tag{5.152}$$

Analyzing the boundary conditions given by Eq. (5.129), it can be concluded that the last two terms of Eq. (5.152) are equal to zero and, therefore, the integral within the first term of this equation should be independent of x. Considering a top-hat profile in $x = 0$ yields the equation

$$\int_0^\infty \overline{\rho}\tilde{u}\tilde{Z} r \, \mathrm{d}r = \frac{\rho_0 u_0 R^2}{2}. \tag{5.153}$$

Substituting Eqs. (5.135), (5.147), and (5.151) into Eq. (5.153) yields an expression for the mixture fraction in the center of the jet line, given by

$$\tilde{Z}_{\mathrm{cl}} = \frac{1}{32}\left(\frac{\rho_0}{\rho_\infty C}\right)\left(\frac{u_0 d}{\nu_t}\right)(\zeta/d)^{-1}(2Sc + 1). \tag{5.154}$$

Thus, an analytical expression for the mixture fraction, corresponding to the product of Eq. (5.154) by Eq. (5.151) is produced as follows

$$\tilde{Z} = \frac{1}{32}\left(\frac{\rho_0}{\rho_\infty C}\right)\left(\frac{u_0 d}{\nu_t}\right)(\zeta/d)^{-1}(2Sc + 1)\left[1 + \frac{(\gamma\eta)^2}{4}\right]^{-2Sc}. \tag{5.155}$$

5.3.4 Analytical Solution for Plumes

The plume, as well as the jet, is a free shear flow. Contrary to the jet, which has some initial momentum when exits a nozzle, the plume movement is governed by buoyancy [26].

Consider a plume, as shown in Figure 5.14, whose main driving force consists of a gradient of density caused by a difference in temperature (natural convection).

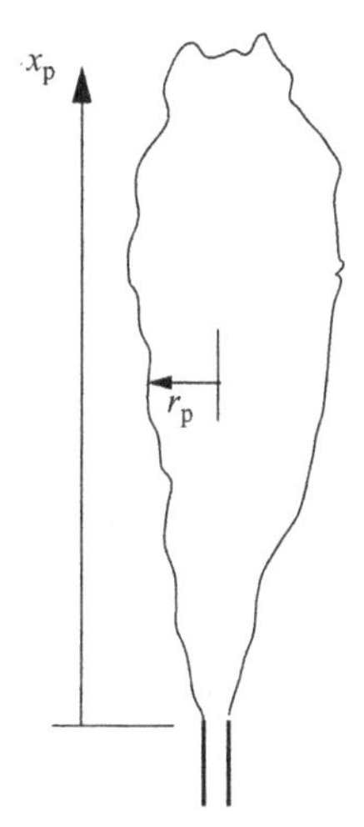

FIGURE 5.14 Schematic representation of a plume.

One writes the expression for the conservation of momentum as:

$$\rho\left(\frac{\partial \vec{v}}{\partial t}+\vec{v}\cdot\vec{\nabla}\vec{v}\right)=-\vec{\nabla}P+\vec{\nabla}\cdot\vec{\vec{\tau}}+\rho\vec{g}. \tag{5.156}$$

In a point distant from the plume, the fluid velocity is virtually zero, so, according to Eq. (5.156), the pressure gradient at this point may be given by

$$\vec{\nabla}P=\rho_{\infty}\vec{g}. \tag{5.157}$$

The velocity gradients in the plume are due to movements caused by differences in the fluid density. Expression (5.157) can be used as an approximation of the gradient of pressure in the system, yielding

$$\rho\left(\frac{\partial \vec{v}}{\partial t}+\vec{v}\cdot\vec{\nabla}\vec{v}\right)=\vec{\nabla}\cdot\vec{\vec{\tau}}+\left(\rho-\rho_{\infty}\right)\vec{g}. \tag{5.158}$$

Similar to the jet case, the governing equations correspond to the equations of continuity, momentum, now with the Boussinesq approximation, and mixture fraction. Writing these expressions in the average of Favre results in

$$\frac{\partial \overline{\rho}}{\partial t}+\frac{\partial\left(\overline{\rho}\tilde{v}_j\right)}{\partial x_j}=0, \tag{5.159}$$

$$\overline{\rho}\left(\frac{\partial \tilde{v}_i}{\partial t}+\tilde{v}_j\frac{\partial \tilde{v}_i}{\partial x_j}\right)=\frac{\partial \overline{\tau}_{j,i}}{\partial x_j}-\frac{\partial\left(\overline{\rho}\widetilde{v_j''v_i''}\right)}{\partial x_j}+\left(\overline{\rho}-\overline{\rho}_{\infty}\right)g_i, \tag{5.160}$$

$$\overline{\rho}\frac{\partial \tilde{Z}}{\partial t} + \overline{\rho}\tilde{v}_j\frac{\partial \tilde{Z}}{\partial x_j} = \overline{\frac{\partial}{\partial x_j}\left(\rho\frac{\nu}{Sc}\frac{\partial Z}{\partial x_j}\right)} - \frac{\partial\left(\overline{\rho}\widetilde{v_j''Z''}\right)}{\partial x_j}. \tag{5.161}$$

In order to simplify the system of equations, the following hypotheses are considered:

1. Steady-state hypothesis, in which time derivatives are zero.
2. Boundary layer approximation, in which second-order derivatives in the axial direction were neglected.
3. The viscous tension $\overline{\tau}_{j,i}$ was neglected when compared to the Reynolds tension $\overline{\rho}\widetilde{v_j''v_i''}$.
4. The molecular component of mixture fraction $\overline{\frac{\partial}{\partial x_j}\left(\rho\frac{\nu}{Sc}\frac{\partial Z}{\partial x_j}\right)}$ was neglected when compared to the turbulent component $\frac{\partial\left(\overline{\rho}\widetilde{v_j''Z''}\right)}{\partial x_j}$.

Similar to the procedure applied in the previous section for a jet, Boussinesq's hypothesis is employed, which considers that the Reynolds stress is proportional to the gradient of velocity. The turbulent mixture fraction, $\overline{\rho}\widetilde{v_j''Z''}$, is modeled as being proportional to the gradient of $\tilde{Z}$ (5.125).

After applying these simplifications, the following equations in cylindrical coordinates are produced:

$$\frac{\partial(\overline{\rho}\tilde{u}r)}{\partial x} + \frac{\partial(\overline{\rho}\tilde{v}r)}{\partial r} = 0, \tag{5.162}$$

$$\overline{\rho}\tilde{u}r\frac{\partial \tilde{u}}{\partial x} + \overline{\rho}\tilde{v}r\frac{\partial \tilde{u}}{\partial r} = \frac{\partial}{\partial r}\left(\overline{\rho}\nu_T r\frac{\partial \tilde{u}}{\partial r}\right) + (\overline{\rho}_\infty - \overline{\rho})\, gr, \tag{5.163}$$

$$\overline{\rho}\tilde{u}r\frac{\partial \tilde{Z}}{\partial x} + \overline{\rho}\tilde{v}r\frac{\partial \tilde{Z}}{\partial r} = \frac{\partial}{\partial r}\left(\frac{\overline{\rho}\nu_T r}{Sc}\frac{\partial \tilde{Z}}{\partial r}\right), \tag{5.164}$$

where $\tilde{u}$ and $\tilde{v}$ correspond to the axial and radial components of velocity, respectively, and $g = -\|\vec{g}\|$, the acceleration due to gravity.

For $r = 0$, the radial velocity component is zero and the mixture fraction is equal to $\tilde{Z}_{\text{cl}}$, while for $r \to \infty$, the axial velocity component and its first order derivative in x are zero, and the mixture fraction is zero due to lack of fuel. Thus, the boundary conditions are given by

$$\begin{aligned} r = 0: &\quad \tilde{v} = 0, \quad \tilde{Z} = \tilde{Z}_{\text{cl}}, \\ r \to \infty: &\quad \tilde{u} = 0, \quad \frac{\partial \tilde{u}}{\partial r} = 0, \quad \tilde{Z} = 0. \end{aligned} \tag{5.165}$$

It is not practical to obtain a closed formula for the axial velocity and the mixture fraction for the plume using the same methodology as for the jet due to the buoyancy term. However, it is possible to determine an estimate for the velocity and the mixture fraction along the plume centerline, considering average values along the radius, according to a top-hat profile. Using these assumptions, the following expressions for average values for continuity, conservation of momentum, and mixture fraction equations are produced:

$$\int_0^\infty \bar{\rho}\tilde{u} r \,\mathrm{d}r = \frac{\rho_\infty \hat{u} b^2}{2}, \tag{5.166}$$

$$\int_0^\infty \bar{\rho}\tilde{u}^2 r \,\mathrm{d}r = \frac{\rho_\infty \hat{u}^2 b^2}{2}, \tag{5.167}$$

$$\int_0^\infty \bar{\rho}\tilde{u}\tilde{Z} r \,\mathrm{d}r = \frac{\rho_\infty \hat{u}\hat{Z} b^2}{2}, \tag{5.168}$$

where $\hat{u}$ and $\hat{Z}$ correspond to the axial component of the velocity vector and the mixture fraction along the radius, respectively, and b is the radius of the plume.

According to Agrawal and Prasad [26], the turbulent viscosity for an axisymmetric plume corresponds to

$$\nu_\mathrm{t} = \frac{U_\mathrm{c} c^2 x}{12\xi^2}\left[5 - 3\exp\left(-\xi^2\right) - 2\exp\left(-\left(C_1^2 - 1\right)\xi^2\right)\right], \tag{5.169}$$

where c is the spreading rate, $\xi = r/(cx)$, $C_1 = c/c_\mathrm{T}$, with c_T the spreading rate for the temperature for a Gaussian distribution along the radius, because

$$\theta = \theta_\mathrm{c}\exp\left(-\frac{r^2}{c_\mathrm{T}^2 x^2}\right). \tag{5.170}$$

U_c corresponds to the axial velocity along the centerline of the plume, which for an axisymmetric plume is proportional to $x^{-1/3}$ [28]. When approaching the exponential term of Eq. (5.169) by a Taylor series truncated at the second term, analogous to the procedure used for the jet, one sees the relation

$$\nu_\mathrm{t} = \frac{U_\mathrm{c} c^2 x}{12}\left(1 + 2C_1^2\right). \tag{5.171}$$

In this case, the eddy viscosity becomes dependent only on x.

In order to obtain expressions for the mean values of $\tilde{u}$, $\tilde{Z}$ and b, we consider the following expressions for axial velocity and mixture fraction:

$$\tilde{u} = \frac{2\nu_\mathrm{t} C\gamma^2}{\zeta}\left(1 + \frac{(\gamma\eta)^2}{4}\right)^{-2}, \tag{5.172}$$

$$\tilde{Z} = \frac{1}{32}\left(\frac{\rho_0}{\rho_\infty C}\right)\left(\frac{u_0 d}{\nu_\mathrm{t}}\right)(\zeta/d)^{-1}(2Sc + 1)\left(1 + \frac{(\gamma\eta)^2}{4}\right)^{-2Sc}, \tag{5.173}$$

obtained using the same methodology applied in Section 5.3.3, neglecting the buoyancy term of the momentum equation (5.163).

Substituting Eqs. (5.172) and (5.173) into Eqs. (5.166), (5.167), and (5.168) yields the average values of the radius, b, the axial velocity, and the plume mixture fraction, according to the equations:

$$b = \frac{2\sqrt{3}\zeta}{\gamma}, \tag{5.174}$$

$$\hat{u} = \frac{2}{3}\frac{\nu_t C\gamma^2}{\zeta}, \tag{5.175}$$

$$\hat{Z} = \frac{\tilde{Z}_{cl}}{2Sc + 1}. \tag{5.176}$$

5.4 MODELS FOR REACTIVE FLOWS IN POROUS MEDIA

This section presents the governing equations for fluid flow in porous media with precipitation reactions, dissolution of minerals, and laminar premixed combustion, as well as similarity parameters. The model is based on Navier-Stokes equations. For modeling precipitation and dissolution, we used the Boussinesq approximation and Darcy's law, which will not be considered in the case of combustion in porous media. Darcy's law, in general, defines the permeability or the ability of a fluid to flow through a porous medium [29]. Another difference from the model of combustion lies in the equations for species, which are based on concentrations.

5.4.1 Model for Reactive Flow Using Darcy's Law

Henry Darcy investigated unidirectional flows in steady-state, as shown in Figure 5.15, through sand beds and found that there is a ratio between the flow rate and the applied pressure difference, which can be written as [30]

$$u = -\frac{K}{\mu}\frac{\partial p}{\partial x}, \tag{5.177}$$

where $\frac{\partial p}{\partial x}$ is the pressure gradient in the flow direction, μ the dynamic viscosity of the fluid, and K the specific permeability, which does not depend on the nature of the fluid, but on the geometry of the medium [30–32].

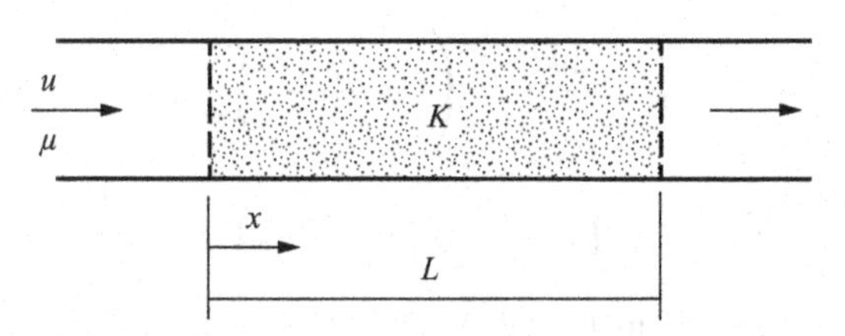

FIGURE 5.15 Flow in porous medium.

In three dimensions, Eq. (5.177) can be written as

$$\vec{v} = -\mu^{-1}\vec{\vec{K}} \cdot \vec{\nabla} p, \tag{5.178}$$

where the permeability $\vec{\vec{K}}$ is a tensor of second order. For the case of an isotropic medium, the permeability is a scalar, and Eq. (5.178) is simplified to

$$\vec{\nabla} p = -\frac{\mu}{K}\vec{v}. \tag{5.179}$$

Permeability may vary, for example, for clean gravel, from 10^{-7} to $10^{-9}\,\mathrm{m}^2$. The Darcy number for permeability is widely used in applications of interest to geophysicists and corresponds to $0.987 \times 10^{-12}\,\mathrm{m}^2$.

Darcy's law is an empirical relationship based on experimental observations of a one-dimensional stream of water through a sand medium [33]. The Darcy equation (5.178) is linear for the flow velocity $\vec{v}$, which is conserved when $\vec{v}$ is sufficiently small. In practice, sufficiently small means that the Reynolds number of the flow, Re_p, based on the pore diameter, is of the order of unity or smaller. As $\vec{v}$ increases, there is no abrupt transition for Re_p from 1 to 10. A change in Eq. (5.179) to consider drag effects gives

$$\vec{\nabla} p = -\frac{\mu}{K}\vec{v} - c_\mathrm{F} K^{-1/2} \rho_\mathrm{f} |\vec{v}|\vec{v}, \tag{5.180}$$

where c_F is a constant drag coefficient in dimensionless form, with a value that varies according to the porosity nature and can be as small as 0.1. Equation (5.180) is known as Forchheimer [30] equation.

Other alternative to the Darcy model is the Brinkman equation. Omitting the inertial terms, it has the form

$$\vec{\nabla} p = -\frac{\mu}{K}\vec{v} + \bar{\mu}\nabla^2\vec{v}. \tag{5.181}$$

This equation contains two viscous terms: The first is Darcy's term and the second term is analogous to the Laplacian that appears in the Navier-Stokes equation. The coefficient $\bar{\mu}$ is the effective viscosity. In averaged form for an isotropic porous medium, $\bar{\mu} = 1/\varphi\tau$, where τ is called the medium tortuosity (whose value ranges from 2.5 to about 50) and $\bar{\mu}/\mu$ depends on the geometry of the medium.

Thus, from the Kozeny-Carman theory, the relationship between porosity and permeability is given by the relation

$$K = \frac{D_\mathrm{p}^2 \phi^3}{180(1-\phi)}, \tag{5.182}$$

where D_p is the characteristic diameter of the grain (material). The constant 180, in Eq. (5.182), has been adjusted experimentally [30]. Hazen [34] determined the following empirical equation for the permeability K

$$K = CD_{\mathrm{p}}^{2} \tag{5.183}$$

and C is a constant.

The Kozeny-Carman relation cannot be used for some materials, such as clay and volcanic compounds. These compounds tend to have large porosities but low permeabilities.

The porosity ϕ is the ratio between the volume of the pores and the total volume. The effective porosity ϕ_{e} corresponds to the ratio between the pore volume available to be occupied by the fluid and the total volume. The effective porosity exclude all pores not connected or that do not contribute to the flow.

Permeability is the measure of the ability of circulating a fluid through a porous medium. Not all porous rocks are permeable, because the pores may be not connected with each other, or be so extremely small that prevent the passage of fluid.

For the migration of hydrocarbons in porous media is generally accepted the Darcy's law. However, Darcy's law is not valid for high speeds, where the inertial forces are not negligible compared with the viscous forces. Therefore, Darcy's law applies only for laminar flows.

To determine the flow regime, the Reynolds number is used

$$Re = \frac{\rho v d}{\mu}, \tag{5.184}$$

where d is the effective pore diameter and v the mean flow velocity.

For straight tubes, the critical Reynolds number, when turbulence starts to appear is of the order of 2000. However, the value of the critical Reynolds number decays considerably when the duct is curved and its diameter varies. To porous media, it is safe to assume that the flow remains laminar while the local Re is less than unity [35].

Another deviation from Darcy's Law occurs for low hydraulic gradients and narrow pores. One possible reason for this anomaly is that the water in the vicinity of the particle surface is subject to adsorption forces and can behave more rigid than a common body of water, exhibiting properties of a Bingham fluid instead of a Newtonian fluid [36].

In the following is presented the set of equations for modeling fluid flow in porous media, where dissolution reactions and/or precipitation of minerals can occur. This set consists of the continuity, momentum, energy, and concentration. The Boussinesq hypothesis is employed to consider the changes in fluid density due to changes in temperature and concentration of the fluid components.

For convection in porous media, many authors use an extension of Eq. (5.179) as follows

$$\rho_{\mathrm{f}}\left[\frac{\partial \vec{v}}{\partial t} + (\vec{v} \cdot \vec{\nabla})\vec{v}\right] = -\vec{\nabla} p - \frac{\mu}{K}\vec{v}. \tag{5.185}$$

5.4.2 Equations for Precipitation/Dissolution of Minerals

Geochemical models have been developed for analysis of chemical reactions in geological systems. These models can be applied in many fields, including environmental protection and remediation, the petroleum industry, and economic geology. Through these models, researchers can understand the composition of natural waters, the mixing and the formation of contaminants in surface water and groundwater, the formation and dissolution of rocks and minerals, among others.

The mineral reactions are in general not in equilibrium, because with the time required for the system to reach the equilibrium state, the reservoir fluid will be already in contact with a different set of minerals. When transportation or other agents change rapidly compared to the reaction rate, the imbalance prevails, and kinetics becomes an important factor when estimating concentrations [37].

The equations that describe the reactive flow can be written as:

1. Continuity

$$\frac{\partial(\phi\rho)}{\partial t} + \vec{\nabla}\cdot(\phi\rho\vec{v}) = 0. \tag{5.186}$$

2. Momentum

$$\frac{\partial(\phi\rho\vec{v})}{\partial t} + \vec{\nabla}\cdot(\phi\rho\vec{v}\vec{v}) = -\phi\vec{\nabla}p + \vec{\nabla}\cdot(\phi\mu\vec{\nabla}\vec{v}) - \frac{\phi\mu\vec{v}}{K} + \phi\rho\vec{g}[\beta_{\mathrm{T}}(T - T_\infty) + \sum_{i=1}^{M}\beta_{C_i}(C_i - C_\infty)]. \tag{5.187}$$

3. Energy (enthalpy)

$$\frac{\partial(\phi\rho h)}{\partial t} + \vec{\nabla}\cdot(\phi\rho\vec{v}h) = +\vec{\nabla}\cdot(\phi D_{\mathrm{T}}\vec{\nabla}T) - h_\nu(T_{\mathrm{S}} - T) + \phi\sum_{i=1}^{N}\dot{w}_i W_i h_i. \tag{5.188}$$

4. Species mass fraction

$$\frac{\partial(\phi\rho Y_i)}{\partial t} + \vec{\nabla}\cdot(\phi\rho\vec{v}Y_i) = +\vec{\nabla}\cdot(\phi D\vec{\nabla}Y_i) \pm \phi\dot{w}_i W_i, \tag{5.189}$$

where

$$\rho = \rho_0\left[1 - \beta_{\mathrm{T}}(T - T_\infty) - \sum_{i=1}^{M}\beta_{C_i}(C_i - C_\infty)\right], \tag{5.190}$$

ϕ is the porosity of the medium, ρ_0 the density of reference, ρ the density of the fluid, $\vec{v}$ the velocity vector, t the time, p the fluid pressure, μ the dynamic viscosity, K the permeability of the medium, g the acceleration due to gravity, T the temperature of the porous medium, c_{P} the specific heat of the medium, h the enthalpy, $C_i = \rho Y_i$ the concentration of species i, D_{T} the coefficient of thermal diffusivity, D the coefficient of mass diffusivity, h_v the coefficient of volumetric transfer of heat between the solid and the gas, $\dot{w}_i$ the reaction rate of the transport equation for the chemical species i, β_{T} the volumetric thermal expansion coefficient, β_{C_i} the coefficient of volumetric expansion due to chemical species i, T_{S} the temperature of the solid matrix, W_i the molecular mass of species i, and M the number of reactions. T_∞ and C_∞ are the temperature and concentration of initial stream, respectively.

The energy equation can be written in terms of the temperature as follows, when neglecting the radiation effects and the diffusive velocity of the i_{th} species

$$\frac{\partial(\phi\rho c_{\mathrm{P}}T)}{\partial t} + \vec{\nabla}\cdot(\phi\rho\vec{v}c_{\mathrm{P}}T) = +\vec{\nabla}\cdot(\phi\kappa_{\mathrm{T}}\vec{\nabla}T) - h_v(T_{\mathrm{S}} - T) - \phi\sum_{i=1}^{N}\dot{w}_iW_ih_i, \tag{5.191}$$

where κ_{T} is the thermal conductivity.

The Poisson equation for pressure comes from the divergence of momentum equation, resulting in

$$\begin{aligned}\nabla^2 p = {}& \frac{1}{\phi}\vec{\nabla}\cdot\left[\vec{\nabla}\cdot(\phi\mu\vec{\nabla}\vec{v}) - \frac{\partial(\phi\rho\vec{v})}{\partial t} - \vec{\nabla}\cdot(\phi\rho\vec{v}\vec{v})\right] \\ & + \vec{\nabla}\cdot\left[-\frac{\mu\vec{v}}{K} + \rho\vec{g}[\beta_{\mathrm{T}}(T - T_\infty) + \sum_{i=1}^{M}\beta_{C_i}(C_i - C_\infty)]\right].\end{aligned} \tag{5.192}$$

As an example, consider the precipitation-dissolution of calcite in an aqueous medium

$$\mathrm{CaCO_3} + \mathrm{H^+} = \mathrm{HCO_3^-} + \mathrm{Ca^{2+}}, \tag{5.193}$$

$$\mathrm{H_2O} + \mathrm{CO_{2(g)}} = \mathrm{H_2CO_3^*}, \tag{5.194}$$

$$\mathrm{H_2CO_3^*} = \mathrm{HCO_3^-} + \mathrm{H^+}, \tag{5.195}$$

$$\mathrm{HCO_3^-} = \mathrm{CO_3^{2-}} + \mathrm{H^+}, \tag{5.196}$$

where $\mathrm{H_2CO_3^*}$ is by convention $\mathrm{H_2CO_3^0} + \mathrm{CO_{2(g)}}$ [38].

For a reaction of the form $A + B \leftrightarrow C$, the dissolution rate is described by [39–41]:

$$\frac{d[A]}{dt} = -k_f[A][B]\left(1 - \frac{[C]}{[A][B]K_{eq}}\right) = k_f[A][B]\left(1 - \frac{IAP}{K_{eq}}\right), \qquad (5.197)$$

where k_f is the rate of the forward reaction, K_{eq} the equilibrium constant, and IAP the product of the activity of the ions.

The $\log\left(\frac{IAP}{K_{eq}}\right)$ is known as the saturation index (Ω) and indicates the state of saturation of the solution with respect to the mineral phase; the situations are [37, 42, 43]: If the IAP is smaller than the equilibrium constant, it indicates a tendency of dissolution, and if IAP is bigger than the equilibrium constant, it indicates a precipitation tendency of the mineral.

For a more detailed analysis, Table 5.1 shows some minerals and equilibrium constants at 25 °C and the activation energy necessary for a kinetic analysis [44].

Some minerals are denominated as feldspars (albite, andesine, anorthite, and k-feldspar), halides (halite), clay minerals (kaolinite), carbonates (calcite, dolomite), and sulfates (anhydrite).

In the Earth crust, the most abundant elements are shown in Table 5.2 with their corresponding percentage in mass.

An iterative method is used for the calculation of activity coefficients. Concentrations of H^+ and OH^- in aqueous solutions are usually very small numbers and therefore are difficult to be handled. Soren Sorensen proposed a more practical measure called pH, which indicates the acidity, neutrality, or alkalinity of an aqueous solution. Mathematically, for ions of H^+, one obtains

$$pH = -\log_{10}[a_{H^+}], \qquad (5.198)$$

where a_{H^+} is the value of the activity of H^+. Thus, the ratio (5.198) may be substituted by

$$pH = -\log_{10}[H^+]. \qquad (5.199)$$

The same idea can be used for ion concentration of OH^-, which is represented by pOH, although this symbol is hardly used. Acidic solutions have pH below 7, alkaline above 7, and neutrality equal to 7.

Atmospheric gases are poorly soluble in water, except the CO_2. The Henry's Law describes the solubility of gases, which establishes a proportionality between the partial pressure of the gas p_i and its concentration. The equilibrium constant can be formulated by replacing the concentration of the gas by the partial pressure. As an example, consider the case

$$CaCO_3 \leftrightarrow CaO + CO_{2(g)}. \qquad (5.200)$$

TABLE 5.1 Equilibrium Constants and Activation Energies for Some Minerals

Mineral	$\log k_e$	E_a (kJ/mol)	Principal Reaction
Quartz	−13.4	90.9	$SiO_{2s} = SiO_{2(aq)}$
Albite	−12.2	69.8	$NaAlSi_3O_8 + 4H^+ = Al^{+++} + Na^+ + 2H_2O + 3SiO_{2(aq)}$
k-Feldspar	−12.4	38.0	$KAlSi_3O_8 + 4H^+ = Al^{+++} + K^+ + 2H_2O + 3SiO_{2(aq)}$
Kaolinite	−13.2	22.2	$Al_2Si_2O_5(OH)_4 + 6H^+ = 2Al^{+++} + 2SiO_{2(aq)} + 5H_2O$
Anorthite	−9.1	17.8	$CaAl_2(SiO_4)_2 + 8H^+ = Ca^{++} + 2Al^{+++} + 2SiO_{2(aq)} + 4H_2O$
Dolomite	−7.5	52.2	$CaMg(CO_3)2 + 2H^+ = Ca^{++} + Mg^{++} + 2HCO_3^-$
Calcite	−5.8	23.5	$CaCO_3 + H^+ = Ca^{++} + HCO_3^-$
Anhydrite	−3.2	31.3	$CaSO_4 = Ca^{++} + SO_4^{--}$
Halite	−0.2	7.4	$NaCl = Na^+ + Cl^-$

TABLE 5.2 Most Abundant Elements in the Earth's Crust

Element Name	Symbol	Percentage
Oxygen	O	46.6
Silicon	Si	27.7
Aluminum	Al	8.1
Iron	Fe	5.0
Calcium	Ca	3.6
Sodium	Na	2.8
Potassium	k	2.6
Magnesium	Mg	2.1

The equilibrium constant can be replaced in this case by the partial pressure of $CO_{2(g)}$, named p_{CO_2},

$$k = \frac{[CaO](p_{CO_2})}{[CaCO_3]}. \tag{5.201}$$

As the activities of CaO and $CaCO_3$ are equal to 1, because they are solids, Eq. (5.201) results in

$$k = p_{CO_2}. \tag{5.202}$$

Therefore, the partial pressure of gas CO_2 is equal to the equilibrium constant in this special case.

The set of equations can be written in dimensionless form as it allows us to compare similar physical phenomena of different length and time scales, and reduces the number of variables in the problem.

5.4.3 Equations of Combustion in Porous Media

Consider a premixed stoichiometric mixture of hydrocarbon gas in a porous media in which the radiation from the gas phase is neglected. The heat transfer is proportional to the temperature difference between the gas and the solid. With these assumptions, the set of governing equations is given by [45–47]

1. Continuity

$$\frac{\partial(\phi\rho)}{\partial t} + \vec{\nabla} \cdot (\phi\rho\vec{v}) = 0. \tag{5.203}$$

2. Momentum

$$\frac{\partial(\phi\rho\vec{v})}{\partial t} + \vec{\nabla}\cdot(\phi\rho\vec{v}\vec{v}) = -\phi\vec{\nabla}p + \vec{\nabla}\cdot(\phi\mu\vec{\nabla}\vec{v}) - \frac{\phi\mu\vec{v}}{K} + \phi\rho\vec{g}[\beta_{\rm T}(T-T_\infty)+\beta_{\rm C}(C-C_\infty)] \quad (5.204)$$

neglecting the inertial and resistance term.

3. Energy (enthalpy)

$$\frac{\partial(\phi\rho_{\rm f}h)}{\partial t} + \vec{\nabla}\cdot(\phi\rho_{\rm f}\vec{v}h) = +\vec{\nabla}\cdot(\phi D_{\rm T}\vec{\nabla}T) - h_v(T_{\rm S}-T) \quad (5.205)$$

$$+\phi\sum_{i=1}^{N}\dot{w}_i W_i h_i. \quad (5.206)$$

4. Temperature for solid matrix

$$\frac{\partial[(1-\phi)\rho_{\rm S}c_{\rm S}T_{\rm S}]}{\partial t} = +\vec{\nabla}\cdot((1-\phi)D_{\rm S}\vec{\nabla}T_{\rm S}) - h_v(T_{\rm S}-T). \quad (5.207)$$

5. Species mass fraction

$$\frac{\partial(\phi\rho Y_i)}{\partial t} + \vec{\nabla}\cdot(\phi\rho\vec{v}Y_i) = +\vec{\nabla}\cdot(\phi D\vec{\nabla}Y_i) + \phi\dot{w}_i W_i, \quad (5.208)$$

where

$$\rho = \rho_0\left[1-\beta_{\rm T}(T-T_\infty) - \rho\sum_{i=1}^{M}\beta_{C_i}(C-C_\infty)\right], \quad (5.209)$$

$\vec{v}$ is the velocity vector, ϕ the porosity, p the pressure, $D_{\rm T}$ the fluid diffusivity, $D_{\rm S}$ the solid diffusivity, D the mass diffusivity, K the permeability, $T_{\rm S}$ the temperature of the solid matrix and T of fluid, ρ the density, h the specific enthalpy of the gas, h_i the heat of combustion, and h_v the coefficient of volumetric transfer of heat between the solid and the gas. Variables $Y_i = \{Y_{\rm F}, Y_{\rm O}, Y_{\rm P}\}^{\rm T}$ represent, respectively, the mass fractions of fuel, oxidant, and combustion products; $c_{\rm P}$ is the specific heat at constant pressure, $\beta_{\rm T}$ and β_{C_i} are the volumetric thermal expansion coefficient and the coefficient of volumetric expansion due to chemical species i, respectively, and $\dot{\omega}_i$ the reaction rate given by $\dot{\omega} = A\rho Y_{\rm F}Y_{\rm O}{\rm e}^{-E_{\rm a}/RT}$, where A is the pre-exponential coefficient, $E_{\rm a}$ the activation energy and R is the gas constant. Concentrations and mass fractions are related by $C_i = \rho Y_i$.

By definition [48], the enthalpy is given by

$$h = \int c_{\rm p}\,{\rm d}T = c_{\rm p}T. \quad (5.210)$$

Thus, replacing Eq. (5.210) in Eq. (5.205), results the energy equation in terms of the temperature of gaseous phase

$$\frac{\partial(\phi\rho_f c_P T)}{\partial t} + \vec{\nabla}\cdot(\phi\rho\vec{v}c_P T) = +\vec{\nabla}\cdot(\phi\kappa_T\vec{\nabla}T) - h_\nu(T_S - T) - \phi\sum_{i=1}^{N}\dot{w}_i W_i h_i, \tag{5.211}$$

where κ_T is the thermal conductivity.

5.4.4 Ionic Dissociation and Precipitation

Ionic dissociation occurs when an ionic compound is added to a solvent, occurring the separation of ions. For example, when adding sodium chloride in water, existing ions in the crystal lattice are separated.

The ionic dissociation is different from ionization. In the dissociation, occurs only separation of the ions that already exist in the crystal before the compound is put into solution. In the ionization there is addition of a molecular compound in a solvent, with breaking of covalent bonds and formation of ions.

The precipitation reaction occurs when anions and cations of an aqueous solution are combined to form a precipitate, for example, the reaction of precipitation of calcium carbonate (calcite):

$$Ca^{2+}_{(aq)} + 2HCO_3{}^-_{(aq)} \rightleftharpoons CaCO_{3\,(s)} + H_2CO_{3\,(aq)}.$$

In this reaction, the calcium ions and bicarbonate are combined in an aqueous medium to form calcium carbonate, which is slightly soluble in water. The calcium carbonate formation in surface waters and its precipitation into the oceanic floor is important in the transference of carbon from surface to deep water [49].

Solubility Rules

The solubility rules assist in predicting the formation of precipitates by ionic reactions, in aqueous medium.

These rules provide guidance in determining which ions will form solids and which will remain in their ionic form in an aqueous solution. The rules should be followed from top to bottom, which means that a compound is soluble (or insoluble) by rule 1, then this statement has precedence over other rules such as the rule 4. The solubility rules are:

1. Salts formed by cations of group I and cations NH_4^+ are soluble. There are some exceptions such as salts formed with Li^+.
2. Acetates $(C_2H_3O_2^-)$, nitrates (NO_3^-) and perchlorates (ClO_4^-) are soluble.
3. Bromides, chlorides, and iodides are soluble.
4. Sulfates $\left(SO_4^{2-}\right)$ are soluble, except sulfates formed with Ca^+, Sr^+, and Ba^{2+}.

5. Salts containing silver $\left(Ag^{+}\right)$, lead $\left(Pb^{+}\right)$, and mercury (I) $\left(Hg^{+}\right)$ are insoluble.
6. Carbonates $\left(CO_2^{2-}\right)$, phosphates $\left(PO_4^{3-}\right)$, sulfides $\left(S^{2-}\right)$, oxides and hydroxides $\left(OH^{-}\right)$ are insoluble. Exceptions are sulfides formed with cations of group II, and hydroxides formed with calcium $\left(Ca^{2+}\right)$, strontium $\left(Sr^{2+}\right)$, and barium $\left(Ba^{2+}\right)$.

REFERENCES

[1] Law CK, Lu T. Towards accommodating realistic fuel chemistry in large-scale computations. In: 46th AIAA aerospace sciences meeting and exhibit, AIAA 2008-969; 2008. p. 1-38.

[2] Pepiot P, Pitsch H. Systematic reduction of large chemical kinetics. In: 4th joint meeting of the U.S. Sections of the Combustion Institute, Philadelphia; 2005.

[3] Nagy T, Turányi T. Reduction of very large reaction mechanism using methods based on simulation error minimization. Combust Flame 2009;156:417-28.

[4] Cormen TH, Leiserson CE, Rivest RL, Stein C. Introduction to algorithms. Cambridge, MA/Boston: MIT Press/McGraw-Hill; 2001.

[5] Kovács T, Zsély I, Kramarics A, Turányi T. Kinetic analysis of mechanism of complex pyrolytic reactions. J Anal Appl Pyrol 2007;79:252-8.

[6] Turányi T. Applications of sensitivity analysis to combustion chemistry. Reliab Eng Syst Saf 1997;57:41-8.

[7] Lebedev AV, Okun MV, Chorkov VA, Tokar PM, Strelkova M. Systematic procedure for reduction of kinetic mechanism of complex chemical processes and its software implementation. J Math Chem 2013;51:73-107.

[8] Warnatz J, Maas U, Dibble RW. Combustion: physical and chemical fundamentals, modeling and simulation, experiments, pollutant formation. Berlin/Heidelberg: Springer-Verlag; 2006.

[9] König K, Maas U. Sensitivity of intrinsic low-dimensional manifolds with respect to kinetic data. Proc Combust Inst 2005;30:1317-23.

[10] König K, Maas U. On-demand generation of reduced mechanism based on hierarchically extended intrinsic low-dimensional manifolds in generalized coordinates. Proc Combust Inst 2009;32:553-60.

[11] Maas U. Efficient calculation of intrinsic low-dimensional manifolds for the simplification of chemical kinetics. Comput Vis Sci 1998;1:69-81.

[12] Bykov V, Maas U. Problem adapted reduced models based on Reaction-Diffusion Manifolds (REDIMs). Proc Combust Inst 2009;32:561-8.

[13] Peters N. Fifteen lectures on laminar and turbulent combustion. Aachen, Germany: Ercoftac Summer School; 1992. Consulted in July 15, 2008, http://www.itv.rwth-aachen.de/fileadmin/LehreSeminar/Combustion/SummerSchool.pdf.

[14] Williams FA. Combustion theory: the fundamental theory of chemically reacting flow systems. Menlo Park, CA: The Bejamin/Cumming Publishing Company Inc; 1985.

[15] Peters N. Turbulent combustion. UK: Cambridge University Press; 2006.

[16] Smith TM, Menon S. Subgrid combustion modeling for premixed turbulent reacting flows, aIAA-1998-0242; 1998.

[17] Pitsch H. A G-equation formulation for large-eddy simulation of premixed turbulent combustion. Stanford, CA: Center for Turbulence Research, Annual Research Briefs; 2002. p. 3-14.
[18] Pitsch H. Improved pollutant predictions in large-eddy simulations of turbulent non-premixed combustion by considering scalar dissipation rate fluctuations. Proc Combust Inst 2002;29:1971-8.
[19] Effelsberg E, Peters N. A composite model for the conserved scalar pdf. Combust Flame 1983;50:351-60.
[20] Poinsot T, Veynante D. Theoretical and numerical combustion. Philadelphia, PA: R. T. Edwards, Inc.; 2001.
[21] Effelsberg E, Peters N. Scalar dissipation rates in turbulent jets and jet diffusion flames. In: Twenty-second symposium on combustion. Pittsburgh, PA: The Combustion Institute; 1988. p. 693-700.
[22] Kim JS, Williams FA. Extinction of diffusion flames with non unity Lewis numbers. J Eng Math 1997;31:101-18.
[23] Gardiner Jr JC. Combustion chemistry. New York: Springer-Verlag; 1984.
[24] Cebeci T. Analysis of turbulent flows. Amsterdam: Elsevier; 2004.
[25] Peters N, Donnerhack S. Structure and similarity of nitric oxide production in turbulent diffusion flames. Symp (Int) Combust 1981;18:33-42.
[26] Agrawal A, Prasad A. Integral solution for the mean flow profiles of turbulent jet, plumes, and wakes. J Fluids Eng 2003;125:813-22.
[27] Schlichting H. Boundary-layer theory. New York: McGraw-Hill Book Company; 1979.
[28] Tennekes H, Lumley JL. A first course in turbulence. Cambridge, MA: MIT Press; 1972.
[29] Dasgupta R, Roy S, Tarafdar S. Correlation between porosity, conductivity and permeability of sedimentary rocks—a ballistic deposition model. Physica A Stat Mech Appl 2000;275:22-32.
[30] Bejan A, Nield DA. Convection in porous media. New York: Springer; 2006.
[31] Chavent G, Jaffré J. Mathematical models and finite elements for reservoir simulation. Amsterdam: North Holland; 1986.
[32] Jakobsen HA. Chemical reactor modeling: multiphase reactive flows. Berlin/Heidelberg: Springer; 2008.
[33] Zeng Z, Grigg R. A criterion for non-Darcy flow in porous media. Transp Porous Media 2006;63:57-69.
[34] Hazen A. Discussion of "Dams on sand foundations" by A. C. Koenig. Trans Am Soc Civil Eng 1911;73:199-203.
[35] Hillel D. Environmental soil physics: fundamentals, applications, and environmental considerations. San Diego, CA: Academic Press; 1998.
[36] Hillel D. Fundamentals of soil physics. New York: Academic Press; 1980.
[37] Kühn M. Reactive flow modeling of hydrothermal systems. In: Lecture notes in Earth sciences. Berlin/Heidelberg: Springer; 2004.
[38] Genthon JP, Röttger H. Reactor dosimetry. Berlin/Heidelberg: Springer; 1985.
[39] Lasaga AC. Chemical kinetics of water-rock interactions. J Geophys Res 1984;89:4009-25.
[40] Geiger C, Kolesov BA. Microscopic-macroscopic relationships in silicates Examples from IR and Raman spectroscopy and heat capacity measurements. In: European notes in mineralogy—energy modeling in minerals, vol. 4; 2002. p. 347-87.
[41] Francisquetti EP. A convective-diffusive-reactive model for fluid migration and combustion in porous media (in Portuguese). PhD thesis. PPGMAp-UFRGS; 2015.
[42] Kehew A. Applied chemical hydrogeology. New Jersey: Prentice Hall; 2001.

[43] Morse JW, Arvidson RS. The dissolution kinetics of major sedimentary carbonate minerals. Earth Sci Rev 2002;58:51-84.
[44] Palandri JL, Kharako YK. A compilation of rate parameters of water mineral interaction kinetics for application to geochemical modeling. US geological survey. Report 2004-1068; 2004.
[45] Barra AJ, Diepvens G, Ellzey JL, Henneke MR. Numerical study of the effects of material properties on flame stabilization in a porous burner. Combust Flame 2003;134:369-79.
[46] Hackert CL, Ellzey JL, Ezekoye OA. Combustion and heat transfer in model two-dimensional porous burners. Combust Flame 1999;116:177-91.
[47] Mendes MAA, Pereira JMC, Pereira JCF. A numerical study of the stability of one-dimensional laminar premixed flames in inert porous media. Combust Flame 2008;153:525-39.
[48] Batchelor GK. An introduction to fluid dynamics. UK: Cambridge University Press; 2000.
[49] Millero FJ. Chemical oceanography. Boca Raton: CRC Press; 2005.

Chapter 6

Numerical Methods for Reactive Flows

The development of rapid and accurate numerical methods has gained prominence in the study of reactive flows. In this chapter, some of the most widely used numerical methods in the solution of such flows are discussed. Real systems have complex geometries for which meshes must be generated.

6.1 CARTESIAN AND GENERALIZED COORDINATES

It is not always convenient to seek solutions in the Cartesian coordinate system. Sometimes, it is necessary to make a coordinate transformation, especially for flows in complex geometries. First, the metrics of the transformation should be determined. Metrics relate the physical plane to the computational plane.

To facilitate comprehension, first consider the transformation of the physical plane (x, y) to the transformed plane (ξ, η) in two dimensions, as shown in Figure 6.1. The total derivatives are given by

$$\mathrm{d}\xi = \xi_x \,\mathrm{d}x + \xi_y \,\mathrm{d}y, \tag{6.1}$$

$$\mathrm{d}\eta = \eta_x \,\mathrm{d}x + \eta_y \,\mathrm{d}y. \tag{6.2}$$

In matrix form,

$$\begin{bmatrix} \mathrm{d}\xi \\ \mathrm{d}\eta \end{bmatrix} = \begin{bmatrix} \xi_x & \xi_y \\ \eta_x & \eta_y \end{bmatrix} \begin{bmatrix} \mathrm{d}x \\ \mathrm{d}y \end{bmatrix} \tag{6.3}$$

or also

$$\left[d^{\mathrm{T}}\right] = [A]\left[d^{\mathrm{P}}\right], \tag{6.4}$$

where $[d^{\mathrm{T}}]$ is the transformed domain and $[d^{\mathrm{P}}]$ is the physical domain. Conversely, it follows that

$$\begin{bmatrix} \mathrm{d}x \\ \mathrm{d}y \end{bmatrix} = \begin{bmatrix} x_\xi & x_\eta \\ y_\xi & y_\eta \end{bmatrix} \begin{bmatrix} \mathrm{d}\xi \\ \mathrm{d}\eta \end{bmatrix} \tag{6.5}$$

Modeling and Simulation of Reactive Flows. http://dx.doi.org/10.1016/B978-0-12-802974-9.00006-4

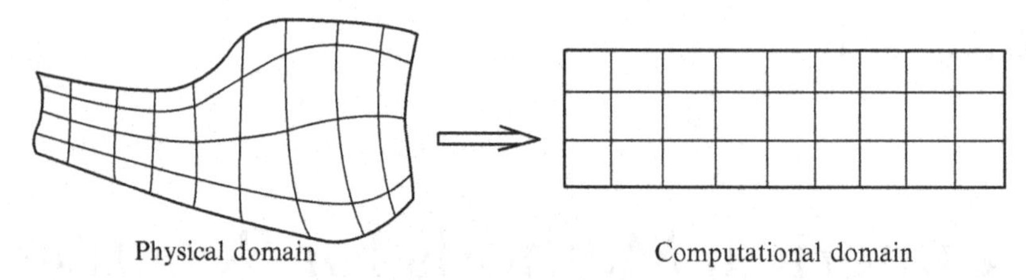

FIGURE 6.1 Physical and computational domains.

or also

$$[d^P] = [B][d^T]. \tag{6.6}$$

Therefore, because $[B^{-1}][d^P] = [d^T] = [A][d^P]$ and $[B^{-1}] = [A]$, gives

$$\xi_x = Jy_\eta, \tag{6.7}$$

$$\xi_y = -Jx_\eta, \tag{6.8}$$

$$\eta_x = -Jy_\xi, \tag{6.9}$$

$$\eta_y = Jx_\xi, \tag{6.10}$$

where the Jacobian is given by

$$J = \frac{1}{\det([B])} = \det\left(\left[B^{-1}\right]\right) = \det([A]). \tag{6.11}$$

The Jacobian is the ratio between the areas of the transformed plane and the physical plane in two dimensions, or the ratio between the corresponding volumes in three dimensions. Thus, the value of the Jacobian must be nonzero.

The metrics of the coordinate transformation, as shown in Figure 6.2, are obtained directly from the figure, as, for example,

$$y_\eta = y(i,j+1) - y(i,j), \tag{6.12}$$

$$x_\eta = x(i,j+1) - x(i,j), \tag{6.13}$$

$$x_\xi = x(i,j) - x(i-1,j), \tag{6.14}$$

$$y_\xi = y(i,j) - y(i-1,j), \tag{6.15}$$

because $\Delta\xi$ and $\Delta\eta$ are assumed to be equal 1 in the transformed plane.

The rule of chain derivation is employed to transform the equations from the Cartesian coordinates to the generalized coordinates. Consider, for example, the transformation of the following equation from Cartesian coordinates to generalized coordinates [1–3]:

$$\frac{\partial \vec{W}}{\partial t} + \frac{\partial \vec{F}_1}{\partial x} + \frac{\partial \vec{F}_2}{\partial y} = 0. \tag{6.16}$$

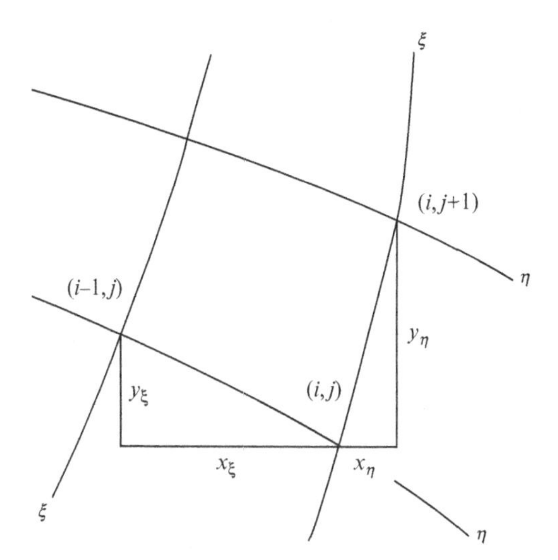

FIGURE 6.2 Metrics of coordinate transformation.

The chain derivative for a generic variable ψ is given by

$$\frac{\partial \vec{\psi}}{\partial x_j} = \frac{\partial \vec{\psi}}{\partial \xi}\frac{\partial \xi}{\partial x_j} + \frac{\partial \vec{\psi}}{\partial \eta}\frac{\partial \eta}{\partial x_j}. \tag{6.17}$$

To transform the system (x, y, t) to the system (ξ, η, τ), one should insert Eq. (6.17) into Eq. (6.16), resulting in

$$\frac{\partial \vec{W}}{\partial \tau} + \frac{\partial \vec{F}_1}{\partial \xi}\frac{\partial \xi}{\partial x} + \frac{\partial \vec{F}_1}{\partial \eta}\frac{\partial \eta}{\partial x} + \frac{\partial \vec{F}_2}{\partial \xi}\frac{\partial \xi}{\partial y} + \frac{\partial \vec{F}_2}{\partial \eta}\frac{\partial \eta}{\partial y} = 0, \tag{6.18}$$

$$\frac{\partial \vec{W}}{\partial \tau} + J\left(\frac{\partial \vec{F}_1}{\partial \xi}\frac{\partial y}{\partial \eta} - \frac{\partial \vec{F}_1}{\partial \eta}\frac{\partial y}{\partial \xi} - \frac{\partial \vec{F}_2}{\partial \xi}\frac{\partial x}{\partial \eta} + \frac{\partial \vec{F}_2}{\partial \eta}\frac{\partial x}{\partial \xi}\right) = 0, \tag{6.19}$$

$$\frac{1}{J}\frac{\partial \vec{W}}{\partial \tau} + \frac{\partial \vec{F}_1}{\partial \xi}\frac{\partial y}{\partial \eta} - \frac{\partial \vec{F}_1}{\partial \eta}\frac{\partial y}{\partial \xi} - \frac{\partial \vec{F}_2}{\partial \xi}\frac{\partial x}{\partial \eta} + \frac{\partial \vec{F}_2}{\partial \eta}\frac{\partial x}{\partial \xi} = 0. \tag{6.20}$$

The latter equation can be rewritten as

$$\begin{aligned} &\frac{1}{J}\frac{\partial \vec{W}}{\partial \tau} + \left[\frac{\partial^2 y}{\partial \xi \partial \eta}\vec{F}_1 + \frac{\partial y}{\partial \eta}\frac{\partial \vec{F}_1}{\partial \xi} - \frac{\partial^2 x}{\partial \xi \partial \eta}\vec{F}_2 - \frac{\partial x}{\partial \eta}\frac{\partial \vec{F}_2}{\partial \xi}\right] \\ &\quad + \left[\frac{\partial^2 x}{\partial \eta \partial \xi}\vec{F}_2 + \frac{\partial x}{\partial \xi}\frac{\partial \vec{F}_2}{\partial \eta} - \frac{\partial^2 y}{\partial \eta \partial \xi}\vec{F}_1 - \frac{\partial y}{\partial \xi}\frac{\partial \vec{F}_1}{\partial \eta}\right] = 0 \end{aligned} \tag{6.21}$$

or,

$$\frac{1}{J}\frac{\partial \vec{W}}{\partial \tau}+\frac{\partial}{\partial \xi}\left[\frac{\partial y}{\partial \eta}\vec{F}_1-\frac{\partial x}{\partial \eta}\vec{F}_2\right]+\frac{\partial}{\partial \eta}\left[\frac{\partial x}{\partial \xi}\vec{F}_2-\frac{\partial y}{\partial \xi}\vec{F}_1\right]=0. \tag{6.22}$$

Thus, Eq. (6.16) in generalized coordinates is shown as

$$\frac{1}{J}\frac{\partial \vec{W}}{\partial \tau}+\frac{\partial \vec{G}_1}{\partial \xi}+\frac{\partial \vec{G}_2}{\partial \eta}=0, \tag{6.23}$$

where the Jacobian is given by the relation (Eq. 6.11), and the terms $\vec{G}_1$ and $\vec{G}_2$ result in, respectively

$$\vec{G}_1 = y_\eta \vec{F}_1 - x_\eta \vec{F}_2, \tag{6.24}$$

$$\vec{G}_2 = -y_\xi \vec{F}_1 + x_\xi \vec{F}_2. \tag{6.25}$$

Consider now the transformation from the physical plane (x, y, z) to the generalized plane (ξ, η, γ) in three dimensions. The total derivatives are given by

$$\mathrm{d}\xi = \xi_x\,\mathrm{d}x + \xi_y\,\mathrm{d}y + \xi_z\,\mathrm{d}z, \tag{6.26}$$

$$\mathrm{d}\eta = \eta_x\,\mathrm{d}x + \eta_y\,\mathrm{d}y + \eta_z\,\mathrm{d}z, \tag{6.27}$$

$$\mathrm{d}\gamma = \gamma_x\,\mathrm{d}x + \gamma_y\,\mathrm{d}y + \gamma_z\,\mathrm{d}z. \tag{6.28}$$

In matrix form, that is

$$\begin{bmatrix}\mathrm{d}\xi\\ \mathrm{d}\eta\\ \mathrm{d}\gamma\end{bmatrix}=\begin{bmatrix}\xi_x & \xi_y & \xi_z\\ \eta_x & \eta_y & \eta_z\\ \gamma_x & \gamma_y & \gamma_z\end{bmatrix}\begin{bmatrix}\mathrm{d}x\\ \mathrm{d}y\\ \mathrm{d}z\end{bmatrix}. \tag{6.29}$$

The physical plane can be written in terms of the transformed plane as

$$\begin{bmatrix}\mathrm{d}x\\ \mathrm{d}y\\ \mathrm{d}z\end{bmatrix}=\begin{bmatrix}x_\xi & x_\eta & x_\gamma\\ y_\xi & y_\eta & y_\gamma\\ z_\xi & z_\eta & z_\gamma\end{bmatrix}\begin{bmatrix}\mathrm{d}\xi\\ \mathrm{d}\eta\\ \mathrm{d}\gamma\end{bmatrix}. \tag{6.30}$$

Solving for the metrics results in

$$\xi_x = J(y_\eta z_\gamma - y_\gamma z_\eta), \tag{6.31}$$

$$\xi_y = J(x_\gamma z_\eta - x_\eta z_\gamma), \tag{6.32}$$

$$\xi_z = J(x_\eta y_\gamma - x_\gamma y_\eta), \tag{6.33}$$

$$\eta_x = J(y_\gamma z_\xi - y_\xi z_\gamma), \tag{6.34}$$

$$\eta_y = J(x_\xi z_\gamma - x_\gamma z_\xi), \tag{6.35}$$

$$\eta_z = J(x_\gamma y_\xi - x_\xi y_\gamma), \tag{6.36}$$

$$\gamma_x = J(y_\gamma z_\eta - y_\eta z_\gamma), \tag{6.37}$$

$$\gamma_y = J(x_\eta z_\gamma - x_\gamma z_\eta), \tag{6.38}$$

$$\gamma_z = J(x_\xi y_\eta - x_\eta y_\xi). \tag{6.39}$$

For the Jacobian, it is convenient to work with the determinant of the matrix

$$J = \det \begin{bmatrix} x_\xi & x_\eta & x_\gamma \\ y_\xi & y_\eta & y_\gamma \\ z_\xi & z_\eta & z_\gamma \end{bmatrix}, \tag{6.40}$$

which results in

$$J = (x_\xi y_\eta z_\gamma + x_\eta y_\gamma z_\xi + x_\gamma y_\xi z_\eta - x_\gamma y_\eta z_\xi - x_\xi y_\gamma z_\eta - x_\eta y_\xi z_\gamma). \tag{6.41}$$

When a non-Cartesian geometry is not too complex, it is convenient to employ Cartesian coordinate for the simulation, using the virtual boundary method.

6.2 VIRTUAL BOUNDARY METHOD

A fluid flowing over a body originates pressure and shear forces on the surface of this body and, in opposition to the flow, the surface exerts opposite forces on the fluid. Thus, a forcing term can be applied to mimic the surface of the body submitted to a flow.

6.2.1 Forcing Term

The introduction of a forcing term $\vec{f}(t, x_s)$ to the momentum equations allows the application of a force field on the surface of a body. Thus, a field of forces can be created from the feedback effect of the velocity field over a body. The forcing term can be calculated as

$$\vec{f}(t, x_s) = \alpha \int_0^t \vec{v}(t^*, x_s)\, \mathrm{d}t^* + \beta \vec{v}(t, x_s), \tag{6.42}$$

where t corresponds to the time, and x_s matches the coordinates of the points along the body surface. In this expression α, of dimension $1/t^2$, determines the frequency of oscillation of the feedback of the velocity field and β, of dimension $1/t$, the damping. For nonstationary flow, α should be large enough so that the resulting natural frequency is greater than the highest frequency present in the flow. So, the force field can produce changes in the flow near the body surface.

Assuming the situation,

$$\frac{\mathrm{d}\vec{v}}{\mathrm{d}t} \sim \vec{f}(t, x_s) = \alpha \int_0^t \vec{v}(t^*, x_s)\, \mathrm{d}t^* + \beta \vec{v}(t, x_s), \tag{6.43}$$

the implication is that, when velocity $\vec{v}$ is different from zero, the forcing term $\vec{f}$ brings $\vec{v}$ back to zero in the surface of the body.

The integral of a variable u in time is approximated by a Riemann sum

$$\int_0^t u(t^*, x_s)\,\mathrm{d}t^* \sim \sum_{j=1}^{N} u(t, x_s)\Delta t, \tag{6.44}$$

where N is the number of time steps and Δt is the size of the time steps.

In the treatment of a body immersed in the flow, it is important to ensure that the external flow is independent of conditions adopted within the body. According to von Terzi et al. [4], the application of the forcing term to generate a virtual body immersed in the flow can be made by using a virtual contour, a solid body, or a counterflow, for example.

The use of the forcing term only in points that define the contour of the submerged body permits fluid to move within the body. When calculating the velocity gradient in the direction normal to the wall by the finite difference method, errors in the prediction of derivative may surge. To solve this problem, the forcing term can be applied to all points within the immersed body in addition to its border points.

Another way of evaluating the virtual contour is by calculating the data using only one side of the surface of the body, which is in contact with the fluid. Theoretically, this approach can maintain the desired accuracy of the numerical scheme in immersed boundaries.

To transfer the information of the body immersed in a flow to adjacent points of the mesh, some techniques can be employed: a Gaussian distribution [5] or a bilinear interpolation [6], for example.

The Gaussian distribution allows smoothing of the transference of forces effect near the surface. Smoothing the field of forces has the effect of spreading the influence of an immersed boundary on the adjacent nodes of the mesh. With the radial distance $d(x)$ of each node adjacent to an immersed contour, one obtains the Gaussian distribution coefficient,

$$\epsilon_s = \mathrm{e}^{-\sigma d^2(x)}, \tag{6.45}$$

where σ is a constant used to adjust the effect of spreading the force field to the vicinity of the immersed boundary. The forcing term is obtained by applying the feedback effect of velocity field in the body. This term is calculated as:

$$\vec{f}(t, x_s) = \epsilon_s \left(\alpha \sum_{t=0}^{t_n} \vec{v}(t, x_s)\Delta t + \beta \vec{v}(t, x_s) \right), \tag{6.46}$$

where n is the number of time steps, $t_n = n\Delta t$ the computational time, and $\vec{v}(t, x_s)$ the characteristic velocity field.

The bilinear interpolation provides the boundary geometric location on the computational mesh, without the need to impose boundary conditions on the nodes located on the interior of the body.

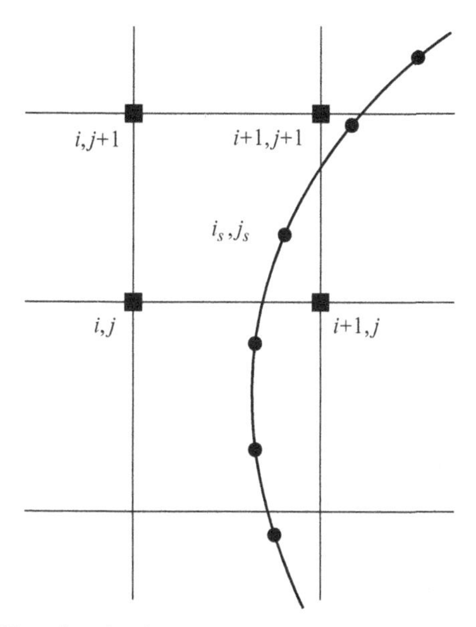

FIGURE 6.3 Virtual boundary sketch.

The number of points, N, should be sufficient to allow proper transfer of information between the immersed body and other mesh points. Four adjacent nodes, $x(i,j)$, $x(i+1,j)$, $x(i,j+1)$, and $x(i+1,j+1)$, exchange information with other mesh points, as shown in Figure 6.3 in two dimensions.

Consider these four adjacent nodes and points on the surface of the body $x_s(i_s,j_s)$, where $i \leq i_s \leq i+1$ and $j \leq j_s \leq j+1$ and $s = 1,2,\ldots,N$. To obtain a linear interpolation coefficient in the longitudinal direction, one writes

$$d(i_s - i) = d(i_s) - d(i) \tag{6.47}$$

or

$$d(i_s - i) = \begin{cases} i_s - (i+1) & i \leq i_s \\ i_s - (i-1) & i > i_s \end{cases} \tag{6.48}$$

and to calculate the linear interpolation coefficient in the transverse direction

$$d(j_s - j) = d(j_s) - d(j). \tag{6.49}$$

Thus, the bilinear interpolation coefficient is given by

$$d_{i,j}(x_s) = d(i_s - i)d(j_s - j). \tag{6.50}$$

The velocity field in the immersed boundary, $\vec{v}(t,x_s)$, is obtained by bilinear interpolation of velocity components of the fluid, $v_{i,j}(t,x_s)$, using the adjacent nodes

$$\vec{v}(t, x_s) = \sum_{i,j}^{i+1,j+1} d_{i,j}(x_s) v_{i,j}(t, x_s). \tag{6.51}$$

Thus, the forcing term due to the velocity field, $\vec{f}_s(t, x_s)$ in each point $x_s(i_s, j_s)$ of the immerse boundary is estimated using

$$\vec{f}_s(t, x_s) = \alpha \sum_{t=0}^{t_n} \vec{v}(t, x_s) \Delta t + \beta \vec{v}(t, x_s). \tag{6.52}$$

6.3 LOW MACH-NUMBER FORMULATION

The Mach number is defined as the ratio of the mean velocity of the fluid by the propagation speed of sound in the medium. According to the Mach number, M, the flow regimes can be classified as incompressible for $M < 0.3$ and compressible for $M > 0.3$ for gases.

Consider the equations of conservation of mass, momentum, and energy. This system of equations has six unknowns: pressure, temperature (or the enthalpy, energy), density, and three velocity components.

When the density does not vary significantly with pressure, the challenge is to determine a pressure field that, when inserted in the equations of momentum, gives rise to a velocity field that satisfies the equation of conservation of mass. The energy equation is used to calculate the temperature and the momentum equations for the calculation of velocity vector components. It is necessary to treat the pressure-velocity coupling once the system is solved separately. Then, after replacing the velocities as a function of the pressure in the conservation of mass equation, the equation for pressure is generated. This formulation is called incompressible formulation. The most important class of problems using this formulation is the heat transfer for low speed flows [3].

If the density varies considerably with the pressure, the state equation relates the density to the pressure and temperature, and the equation of conservation of mass is used for the density. This formulation, in which all dependent variables have their evolution equation, is called compressible formulation. This formulation is mainly employed for high speed flows, where the main interest is the determination of the pressure coefficient on the surface of a body to calculate the drag forces and aerodynamic coefficients.

Combustion and geochemistry characteristic velocities are usually small compared with the speed of sound. As the Mach number approches zero, the contribution of the pressure gradient in the nondimensional momentum equations, $\nabla p/M^2$, becomes singular. So, a numerical method used to integrate the original set of equations tends to fail when applied to very low Mach numbers in combustion [7] and geochemistry. Therefore, one needs a numerical technique that solves the original compressible flow equations, but that can also be efficiently used at low Mach-numbers.

The common formulations of variable density for low Mach numbers are those used when acoustic waves are filtered. Density will change due to temperature variations, but remains independent from the pressure. Then, one can decouple the pressure field of the state equation from the pressure gradients of the momentum equations. With the elimination of acoustic waves, a larger time-step of this formulation can be chosen than when using the standard compressible formulation [8]; however, the obtained formulation is not adequate to analyze combustion instabilities, for example.

The popular numerical approaches used for low Mach-number flows are

- SIMPLE type algorithms [9].
- projection type methods [10].

The semi-implicit method for pressure-linked equations (SIMPLE) was initially developed for the simulation of incompressible flows [9]. The strategy is to write an iteractive scheme whose final result is a pressure correction that is applied to update the velocities at time level t^{n+1} based on time level t^n. This method was extended to solve compressible flows. The idea was to introduce a thermodynamic coupling between the pressure and the density using the mass conservation equation to obtain the pressure correction [11].

Projection-type methods were also developed to solve incompressible flows [12]. A projection scheme consists of two steps: first the momentum equations are solved without the pressure term for one time-step; this gives the velocities. The second step consists of projecting the intermediate velocity back onto the space of the divergence free field (for more details see Ferziger and Péric [12]).

In a reactive flow with heat release, the dynamics of the fluid motion is coupled with the chemical reaction through the nonhomogeneous distribution due to the thermal expansion. Such a dynamic process must be solved by an appropriate formulation. Fully compressible flow descriptions are the ones where density and pressure vary. They are more appropriate for flame-acoustic interactions and combustion in high-speed flows, but their computational costs are high at low-speed flow [13].

Many direct numerical simulations (DNS) for combustion flows were developed considering constant density [13]. Constant density increases the stability of a numerical code considerably. But, these formulations are limited by the absence of flame-induced flow modifications due to heat release. On the other hand, if the fully compressible equations are employed to solve low Mach number flows, the high-frequency acoustic waves create severe restrictions on the time-stepping increment.

For a very low Mach number, the round-off error can contaminate the pressure gradient calculation, and the turbulent flow can present errors [8]. The main assumption of the low Mach-number formulation is that the pressure can be expanded as $p = p_0(t) + \gamma M^2 p_1$. To calculate the pressure and velocities, first the momentum equations can be calculated without the pressure gradient, then the pressure can be estimated by a Poisson's equation, and finally the velocities can be corrected. The density is obtained from the state equation for p_0 equal to a constant ($p_0 = 1$ atm).

The formulation for a low Mach number is usually applied to combustion problems. An increase in thermal energy results in variations of pressure and temperature, gas expansion, wave propagation, and generation of nontrivial flows carrying kinetic energy [7].

In this formulation, the density comes from the state relation for $p \sim p_0$, which means that

$$\rho \sim \frac{p_0}{RT} \tag{6.53}$$

and may be relaxed to avoid numerical instability due to rapid variations of temperature, using

$$\rho \sim \frac{p_0}{\frac{\alpha T}{1-\alpha} + 1} \tag{6.54}$$

with the relaxation coefficient α ranging from 0.1 to 0.8.

6.4 LARGE-EDDY SIMULATION (LES)

The LES is a technique employed to solve turbulent flows. In the LES methodology, large-scale eddies, which are dependent on the geometry of the flow, are computed explicitly, while small eddies, which have more universal behavior, are implicitly approximated by sub-grid models [14].

DNS solves the complete set of equations with no need of models for turbulent viscosity. LES needs subgrid models, but less refined meshes than DNS; it can be used with higher Reynolds numbers (realistic) than in DNS. The quality and accuracy of the results are directly related to these sub-grid models. LES is a promising approach for the study of combustion because the large structures in turbulent flows are generally dependent on the geometry of the system. Compared to Reynolds averaged Navier-Stokes (RANS) simulations, LES simulations are more expensive and they always are three-dimensional simulations.

Basically, there are four conceptual steps in LES:

1. The filtering operation is performed by decomposing the velocity u in a filtered component $\tilde{u}$, representative of large scales, and a variation (sub-grid) component u'.
2. The governing equations for the velocity field are derived from the filtered Navier-Stokes equations.
3. The closure of the system of equations is obtained by modeling the residual stress tensor, using a model for the turbulent viscosity.
4. The filtered equations are solved numerically for $\tilde{u}$, which provides an approach to large scales.

In general, the filtering operation in LES is made through the convolution of the variable of interest with the filter function G. The filter can be applied in space, time or in both as

$$\overline{u}(x,t) = \int_{-\infty}^{\infty}\int_{-\infty}^{\infty} G(r,\tau)u(x-r,t-\tau)\ dr\ d\tau. \tag{6.55}$$

The filter function G contains scales of length and time denoted by Δ and τ_c, respectively. Scales smaller than Δ and τ_c are eliminated when applying the filter.

The variable u is decomposed into a filtered component and a fluctuation, according to

$$u = \overline{u} + u'. \tag{6.56}$$

The filtering procedure may appear similar to the application of the Reynolds average; however, contrary to the latter, the residual component is a random field, so that the filtering is not zero, or

$$\overline{u'} \neq 0. \tag{6.57}$$

To facilitate the manipulation of the Navier-Stokes equations, it is necessary that the filter satisfies the following [15]:

1. Conservation of constants: Filtering a constant variable should return the same constant $\left(\bar{k} = k\right)$,

$$\int_{-\infty}^{\infty}\int_{-\infty}^{\infty} G(r,\tau)\,dr\,d\tau = 1. \tag{6.58}$$

2. Linearity

$$\overline{a+b} = \overline{a} + \overline{b}. \tag{6.59}$$

3. Computation of derivative

$$\overline{\frac{\partial u}{\partial x_j}} = \frac{\partial \overline{u}}{\partial x_j}. \tag{6.60}$$

Some commonly filters used in LES are the cut-off, box, and Gaussian. For flows with density, ρ, variable, the Favre filter is given by

$$\overline{\rho}\tilde{f}(\mathbf{x}) = \int \rho\tilde{f}(\mathbf{x}')F(\mathbf{x}-\mathbf{x}')\,d\mathbf{x}'. \tag{6.61}$$

To obtain the equations in the averages of Reynolds and Favre (Section 4.2.2), models for the turbulent viscosity are necessary. The models commonly used in LES consist of the Smagorinsky and the Germano. So, we prefer to write the equations in Favre form and use as a filter the computational mesh.

6.5 METHODS OF DISCRETIZATION OF FLOW EQUATIONS

The principal objective of numerical methods is to solve even more complex flows, to preserve the properties of the flow locally, and to save computational time. Among the most widely used methods to solve flows, there are the methods of finite difference, finite volume, and finite element. Each of these methods has its own advantages and disadvantages, which are well discussed in the literature. It is understood that, for reactive flows of technical interest, an approximation of second order in space and in time is frequently sufficient.

6.5.1 Finite Difference Method

The finite difference method is widely used in computational fluid dynamics. It produces a system that can be solved for the variables of flow in the computational mesh. Furthermore, it is easy to implement and produces satisfactory results for reactive flows.

Multistep methods can be used for integration of the system of equations in time, such as Runge-Kutta, and the central finite difference scheme for the spatial approximation of first-order derivatives for each grid point (i, j, k) as shown in Figure 6.4 for uniform meshes by simplicity

$$\left(\frac{\partial f}{\partial x}\right)_{(i,j,k)} = \frac{f_{(i+1,j,k)} - f_{(i-1,j,k)}}{2\Delta x} + O((\Delta x)^2), \tag{6.62}$$

$$\left(\frac{\partial f}{\partial y}\right)_{(i,j,k)} = \frac{f_{(i,j+1,k)} - f_{(i,j-1,k)}}{2\Delta y} + O((\Delta y)^2), \tag{6.63}$$

$$\left(\frac{\partial f}{\partial z}\right)_{(i,j,k)} = \frac{f_{(i,j,k+1)} - f_{(i,j,k-1)}}{2\Delta z} + O((\Delta z)^2). \tag{6.64}$$

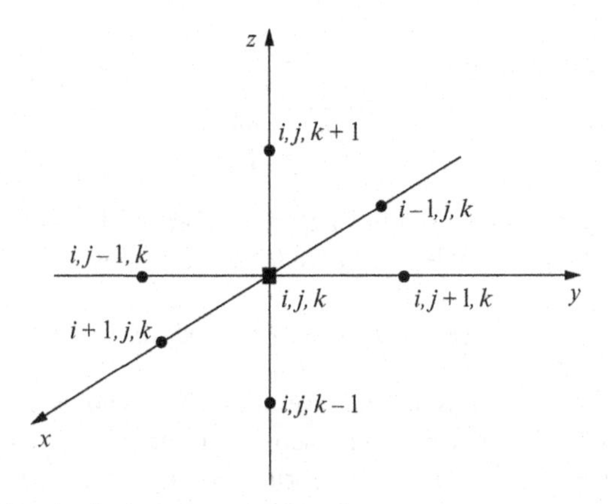

FIGURE 6.4 Grid points in the computational mesh.

For variables with large variations, such as the mixture fraction, it is convenient to use a flux limiter, ensuring that the total variation of the mixture fraction will not grow with time. The idea is to limit the spatial derivatives to realistic values. The flux limiters avoid spurious oscillations occurring in the high-order schemes due to shocks, discontinuities, or rapid changes in solution, acting only where the discontinuities are present. Using, for example, a total variation diminishing (TVD) [16, 17] scheme, the first-order derivative in the x-direction, is given by

$$\left(\frac{\partial f}{\partial x}\right)_{(i,j,k)} \sim f^{\text{low}}_{(i,j,k)} - \phi(r_i)\left[f^{\text{low}}_{(i,j,k)} - f^{\text{high}}_{(i,j,k)}\right], \tag{6.65}$$

where

$$f^{\text{low}}_{(i,j,k)} = \frac{f_{(i,j,k)} - f_{(i-1,j,k)}}{\Delta x} + O(\Delta x), \tag{6.66}$$

$$f^{\text{high}}_{(i,j,k)} = \frac{f_{(i+1,j,k)} - f_{(i-1,j,k)}}{2\Delta x} + O((\Delta x)^2), \tag{6.67}$$

and the flux limiter ϕ of Van Leer is defined as [18]

$$\phi(r_i) = \frac{r_i + |r_i|}{1 + r_i} \tag{6.68}$$

with

$$r_i = \frac{f_{(i,j,k)} - f_{(i-1,j,k)}}{f_{(i+1,j,k)} - f_{(i,j,k)}}, \qquad \lim_{r_i \to \infty} \phi(r_i) = 2. \tag{6.69}$$

The approximation given in Eq. (6.65) is of second order, being of first order in regions of discontinuities.

The second-order derivatives in space can be approximated with central differences as

$$\left(\frac{\partial^2 f}{\partial x^2}\right)_{(i,j,k)} = \frac{f_{(i+1,j,k)} - 2f_{(i,j,k)} + f_{(i-1,j,k)}}{(\Delta x)^2} + O((\Delta x)^2), \tag{6.70}$$

$$\left(\frac{\partial^2 f}{\partial y^2}\right)_{(i,j,k)} = \frac{f_{(i,j+1,k)} - 2f_{(i,j,k)} + f_{(i,j-1,k)}}{(\Delta y)^2} + O((\Delta y)^2), \tag{6.71}$$

$$\left(\frac{\partial^2 f}{\partial z^2}\right)_{(i,j,k)} = \frac{f_{(i,j,k+1)} - 2f_{(i,j,k)} + f_{(i,j,k-1)}}{(\Delta z)^2} + O((\Delta z)^2), \tag{6.72}$$

where

$$\Delta x = \frac{x_{i+1} - x_{i-1}}{2}, \tag{6.73}$$

$$\Delta y = \frac{y_{j+1} - y_{j-1}}{2}, \tag{6.74}$$

$$\Delta z = \frac{z_{k+1} - z_{k-1}}{2}. \tag{6.75}$$

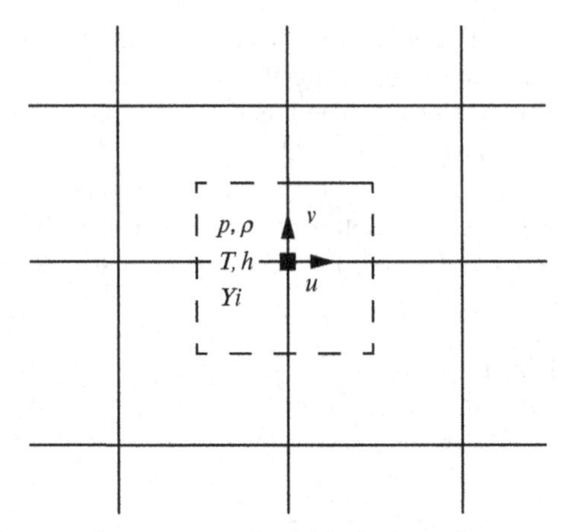

FIGURE 6.5 Node-centered arrangement of variables in the mesh.

In the following is shown the discretization for a model equation in two dimensions, given by

$$\frac{\partial \phi}{\partial t} = \nu \left(\frac{\partial^2 \phi}{\partial x^2} + \frac{\partial^2 \phi}{\partial y^2} \right) - \left(\frac{\partial (u\phi)}{\partial x} + \frac{\partial (v\phi)}{\partial y} \right) + f. \tag{6.76}$$

Model Eq. (6.76) can be discretized in space using a central finite difference scheme of the second order, for a node-centered mesh, as shown in Figure 6.5, as

$$\begin{aligned} \left(\frac{\partial \phi}{\partial t} \right)_{i,j} = \nu & \left(\frac{\phi_{i+1,j} - 2\phi_{i,j} + \phi_{i-1,j}}{\Delta x^2} + \frac{\phi_{i,j+1} - 2\phi_{i,j} + \phi_{i,j-1}}{\Delta y^2} \right) \\ & - \left(\frac{(u\phi)_{i+1,j} - (u\phi)_{i-1,j}}{2\Delta x} + \frac{(v\phi)_{i,j+1} - (v\phi)_{i,j-1}}{2\Delta y} \right) + f_{i,j}, \end{aligned} \tag{6.77}$$

where $f_{i,j}$ is the finite difference approximation of second order of the source term.

6.5.2 Finite Volume Method

The finite volume method is frequently employed in commercial codes for solving heat transfer and computational fluid dynamics.

Consider the set of flow equations in Cartesian coordinates. Integrating it in the control volume and using the divergence theorem (Eq. A.32) gives

$$\int_\Omega \frac{\partial \phi}{\partial t} \, d\Omega + \int_S \left(\bar{\bar{G}} \cdot \vec{n} \right) dS = 0, \tag{6.78}$$

where $\bar{\bar{G}}$ is the tensor of convective and diffusive fluxes. For a given control volume (i, j, k), this integration results in

$$\left(\frac{\mathrm{d}}{\mathrm{d}t}\phi_{i,j,k}\right) + \vec{Q}_{i,j,k} = 0, \tag{6.79}$$

$$\begin{aligned}\vec{Q}_{i,j,k} = &\frac{1}{\Omega_{i,j,k}}\Big(\bar{\bar{G}}_{i,j+1/2,k}\cdot\vec{S}_{i,j+1/2,k} - \bar{\bar{G}}_{i,j-1/2,k}\cdot\vec{S}_{i,j-1/2,k}\\ &+\bar{\bar{G}}_{i+1/2,j,k}\cdot\vec{S}_{i+1/2,j,k} - \bar{\bar{G}}_{i-1/2,j,k}\cdot\vec{S}_{i-1/2,j,k}\\ &+\bar{\bar{G}}_{i,j,k+1/2}\cdot\vec{S}_{i,j,k+1/2} - \bar{\bar{G}}_{i,j,k-1/2}\cdot\vec{S}_{i,j,k-1/2}\Big),\end{aligned} \tag{6.80}$$

where $\Omega_{i,j,k}$ is the volume of the cell (i, j, k).

The values of the tensor $\bar{\bar{G}}_{i,j+1/2,k}$ are associated to the center of the north side of the cell (i, j, k), and are calculated by interpolation between the values of ϕ at the center of the cell (i, j, k) and $(i, j + 1, k)$. The vector $\vec{S}_{i,j+1/2,k}$ is normal to the north side of the cell (i, j, k).

Because the variables are stored only in the centers of the control volumes for computational cost savings, it is necessary to estimate their values through averages on the cell interfaces. The values of the flow rates, at the centers of the faces of the control volume, can be approximated by taking a simple average between the values of two cells, to find the value in the middle of the face that separate them, for example,

$$\bar{\bar{G}}_{i,j+1/2,k} = \bar{\bar{G}}\left[\frac{1}{2}\left(\vec{W}_{i,j,k} + \vec{W}_{i,j+1,k}\right)\right]. \tag{6.81}$$

The finite volume method based on central differences is not dissipative. Thus, high-frequency oscillations of error are not damped near discontinuities, and the procedure cannot converge to the solution in steady state. To eliminate these spurious oscillations, dissipative terms can be introduced explicitly through artificial dissipation [1, 19].

For example, for the model problem illustrated by Eq. (6.76), one obtains

$$\begin{aligned}\left(\frac{\partial\phi}{\partial t}\right)_{i,j} &= \nu\left(\frac{\phi_{i+1,j} - 2\phi_{i,j} + \phi_{i-1,j}}{\Delta x^2} + \frac{\phi_{i,j+1} - 2\phi_{i,j} + \phi_{i,j-1}}{\Delta y^2}\right)\\ &- \left(\frac{(u\phi)_{i+1/2,j} - (u\phi)_{i-1/2,j}}{\Delta x} + \frac{(v\phi)_{i,j+1/2} - (v\phi)_{i,j-1/2}}{\Delta y}\right) + f_{i,j},\end{aligned} \tag{6.82}$$

based on the node-centered arrangement, for an uniform mesh as shown in Figure 6.5, or in the cell-centered arrangement, as shown in Figure 6.7.

Depending on the formulation and how the pressure equation is written for compressible or incompressible flows, many forms are available to approximate ϕ in each face of control volume to perform the mass balance in each cell [9, 12].

In finite volume, cell-centered, node-centered, and staggered arrangements are frequently used, as shown in Figures 6.5, 6.6, and 6.7. Each of these arrangements has its own advantages and disadvantages, which are well discussed in the literature.

The tendency seems to be to use the collocated arrangement (cell centered and node-centered) of the flow variables in the mesh, due to the decrease of computational memory, especially for three-dimensional flow situations. Besides, it seems to be more consistent to evaluate all unknowns in the same control volume.

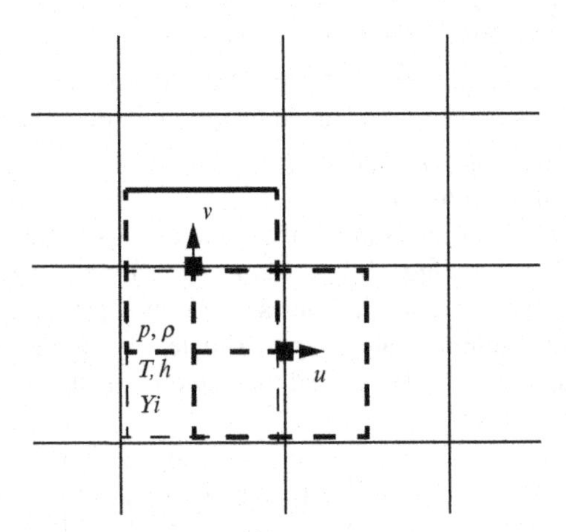

FIGURE 6.6 Staggered arrangement of variables in the mesh.

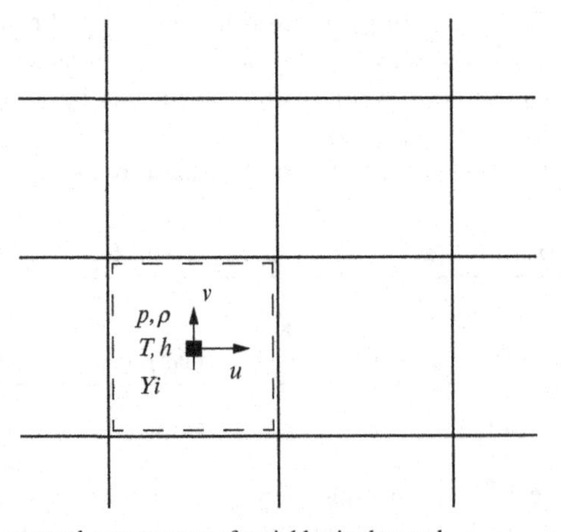

FIGURE 6.7 Cell-centered arrangement of variables in the mesh.

6.5.3 Finite Element Method

When using the finite element method, the interpolation functions are written in the form

$$\phi = \sum_{i=1}^{n} N_i \phi_i. \tag{6.83}$$

For a general convective problem in three dimensions,

$$\frac{\partial \phi}{\partial t} = \nu \nabla^2 \phi - \vec{v}.\vec{\nabla}\phi + f, \tag{6.84}$$

after integration by parts, in a domain Ω with boundary S, results in

$$[K_t]\left\{\frac{\partial \phi}{\partial t}\right\} = [K]\{\phi\} - [C]\{\phi\} + [K_s]\{\phi\} + \{F\}, \tag{6.85}$$

where

$$K_{t\,i,j} = \int_\Omega N_i N_j \, \mathrm{d}\Omega, \tag{6.86}$$

$$K_{i,j} = \nu \int_\Omega \left(\frac{\partial N_i}{\partial x}\frac{\partial N_j}{\partial x} + \frac{\partial N_i}{\partial y}\frac{\partial N_j}{\partial y} + \frac{\partial N_i}{\partial z}\frac{\partial N_j}{\partial z} \right) \mathrm{d}\Omega, \tag{6.87}$$

$$C_{i,j} = \int_\Omega N_i \left(\frac{\partial (uN_j)}{\partial x} + \frac{\partial (vN_j)}{\partial y} + \frac{\partial (wN_j)}{\partial z} \right) \mathrm{d}\Omega, \tag{6.88}$$

$$K_{s\,i,j} = \int_S h \, N_i N_j \, \mathrm{d}S, \tag{6.89}$$

$$F_i = \int_\Omega f \, N_i \, \mathrm{d}\Omega + \int_S q_i \, \mathrm{d}S, \tag{6.90}$$

where (i,j) refers here to each coefficient of the local finite element matrix.

For three-dimensional flow using a tetrahedral element, as shown in Figure 6.8, the local natural coordinates and the global coordinates are related by

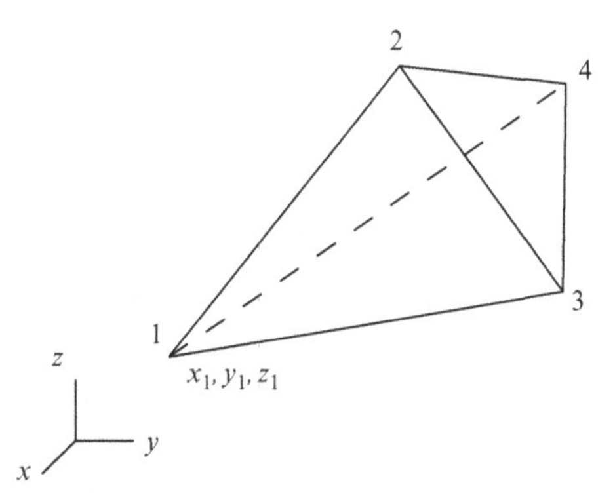

FIGURE 6.8 Tetrahedral element used in finite element.

$$
\begin{aligned}
x &= L_1x_1 + L_2x_2 + L_3x_3 + L_4x_4, \\
y &= L_1y_1 + L_2y_2 + L_3y_3 + L_4y_4, \\
z &= L_1z_1 + L_2z_2 + L_3z_3 + L_4z_4, \\
1 &= L_1 + L_2 + L_3 + L_4.
\end{aligned}
\tag{6.91}
$$

Solving this system yields

$$N_i = L_i = \frac{1}{6V}(a_i + b_ix + c_iy + d_iz), \tag{6.92}$$

where

$$6V = \begin{bmatrix} 1 & x_1 & y_1 & z_1 \\ 1 & x_2 & y_2 & z_2 \\ 1 & x_3 & y_3 & z_3 \\ 1 & x_4 & y_4 & z_4 \end{bmatrix} \tag{6.93}$$

and

$$a_1 = \begin{bmatrix} x_2 & y_2 & z_2 \\ x_3 & y_3 & z_3 \\ x_4 & y_4 & z_4 \end{bmatrix} \tag{6.94}$$

$$b_1 = -\begin{bmatrix} 1 & y_2 & z_2 \\ 1 & y_3 & z_3 \\ 1 & y_4 & z_4 \end{bmatrix} \tag{6.95}$$

$$c_1 = -\begin{bmatrix} x_2 & 1 & z_2 \\ x_3 & 1 & z_3 \\ x_4 & 1 & z_4 \end{bmatrix} \tag{6.96}$$

$$d_1 = -\begin{bmatrix} x_2 & y_2 & 1 \\ x_3 & y_3 & 1 \\ x_4 & y_4 & 1 \end{bmatrix}. \tag{6.97}$$

Other constants are obtained in a similar manner.

For a bidimensional flow, use a triangular element in natural coordinates, as shown in Figure 6.9 to obtain the system

$$
\begin{aligned}
x &= L_1x_1 + L_2x_2 + L_3x_3, \\
y &= L_1y_1 + L_2y_2 + L_3y_3, \\
1 &= L_1 + L_2 + L_3.
\end{aligned}
\tag{6.98}
$$

In this way, the bilinear interpolation yields

$$N_i = L_i = \frac{1}{2A}(a_i + b_ix + c_iy), \tag{6.99}$$

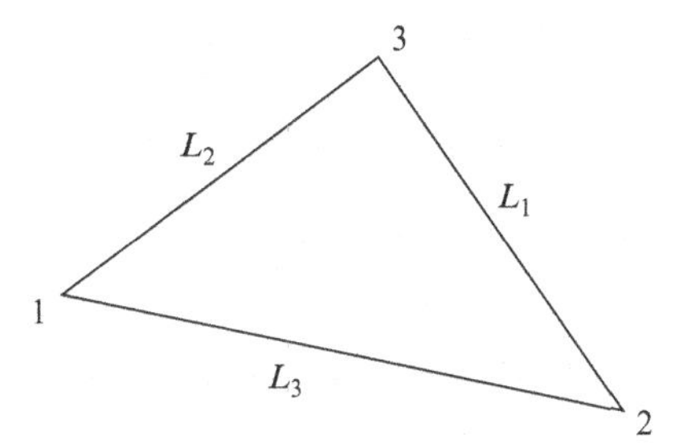

FIGURE 6.9 Triangular element frequently used in finite element.

that is,

$$N_1 = L_1 = \frac{1}{2A}(a_1 + b_1 x + c_1 y), \tag{6.100}$$

$$N_2 = L_2 = \frac{1}{2A}(a_2 + b_2 x + c_2 y), \tag{6.101}$$

$$N_3 = L_3 = \frac{1}{2A}(a_3 + b_3 x + c_3 y), \tag{6.102}$$

where

$$2A = \begin{bmatrix} 1 & x_1 & y_1 \\ 1 & x_2 & y_2 \\ 1 & x_3 & y_3 \end{bmatrix} \tag{6.103}$$

and

$$\begin{aligned} a_1 &= x_2 y_3 - x_3 y_2, \\ a_2 &= x_3 y_1 - x_1 y_3, \\ a_3 &= x_1 y_2 - x_2 y_1, \\ b_1 &= y_2 - y_3, \\ b_2 &= y_3 - y_1, \\ b_3 &= y_1 - y_2, \\ c_1 &= x_3 - x_2, \\ c_2 &= x_1 - x_3, \end{aligned} \tag{6.104}$$

$$c_3 = x_2 - x_1. \tag{6.105}$$

The matrix system for each element is assembled, as given by Eq. (6.85), and the matrix for the overall system is obtained via superposition of all elements of the domain, using the connectivities among them. Triangular elements are used most often in the finite element if the geometry is complex.

Triangular elements can also be used in finite difference or finite volume without difficulties. However, a rectangular element can be more easily applied in finite difference or in finite volume than in finite element. The resulting system can be solved by explicit or implicit methods; the presented form is more suitable for solution via methods of Runge-Kutta family.

6.6 HIGH-ORDER SCHEMES AND COMPACT SCHEMES

The derivatives that appear more frequently in the equations of reactive flows are of the first and second orders. Conventional approaches to these derivatives are presented in the following. It is important not to confuse order of derivative with order of its approximation.

The approximations of first and second orders for the derivative of first order are [20, 21]:

First-order approximations for derivative of first order:

$$y_i' = \frac{1}{h}(y_{i+1} - y_i) + O(h) \qquad \text{forward} \tag{6.106}$$

$$y_i' = \frac{1}{h}(y_i - y_{i-1}) + O(h) \qquad \text{backward.} \tag{6.107}$$

Second-order approximations for the derivative of first order:

$$y_i' = \frac{1}{2h}(y_{i+1} - y_{i-1}) + O(h^2) \qquad \text{centered} \tag{6.108}$$

$$y_i' = \frac{1}{2h}(-3y_i + 4y_{i+1} - y_{i+2}) + O(h^2) \qquad \text{forward} \tag{6.109}$$

$$y_i' = \frac{1}{2h}(3y_i - 4y_{i-1} + y_{i-2}) + O(h^2) \qquad \text{backward.} \tag{6.110}$$

Second-order approximations for the derivative of second order:

$$y_i'' = \frac{1}{h^2}(y_i - 2y_{i+1} + y_{i+2}) + O(h^2) \qquad \text{forward} \tag{6.111}$$

$$y_i'' = \frac{1}{h^2}(y_i - 2y_{i-1} + y_{i-2}) + O(h^2) \qquad \text{backward} \tag{6.112}$$

$$y_i'' = \frac{1}{h^2}(y_{i+1} - 2y_i + y_{i-1}) + O(h^2) \qquad \text{centered.} \tag{6.113}$$

Often it is most appropriate to increase the spatial order of approximation than to increase the number of grid points. An increase in one order of the method would be approximately comparable to an increase in the points of the mesh by an order of magnitude in each coordinate direction. The higher order approximations usually go up to the order 4. Accordingly, in the following, these approximations are presented for the derivatives of first and second orders [21, 22].

Third-order approximations for the derivative of first order:

$$y'_i = \frac{1}{6h}(2y_{i+3} - 9y_{i+2} + 18y_{i+1} - 11y_i) + O(h^3) \qquad \text{forward} \qquad (6.114)$$

$$y'_i = \frac{1}{6h}(11y_i - 18y_{i-1} + 9y_{i-2} - 2y_{i-3}) + O(h^3) \qquad \text{backward.} \qquad (6.115)$$

Fourth-order approximations for the derivative of first order:

$$y'_i = \frac{1}{12h}(-3y_{i+4} + 16y_{i+3} - 36y_{i+2} + 48y_{i+1} - 25y_i) + O(h^4) \qquad \text{forward} \qquad (6.116)$$

$$y'_i = \frac{1}{12h}(25y_i - 48y_{i-1} + 36y_{i-2} - 16y_{i-3} + 3y_{i-4}) + O(h^4) \qquad \text{backward} \qquad (6.117)$$

$$y'_i = \frac{1}{12h}(-y_{i+2} + 8y_{i+1} - 8y_{i-1} + y_{i-2}) + O(h^4) \qquad \text{centered.} \qquad (6.118)$$

Fourth-order approximations for the derivative of second order:

$$y''_i = \frac{1}{12h^2}(11y_{i+4} - 56y_{i+3} + 114y_{i+2} - 104y_{i+1} + 35y_i) + O(h^4) \qquad \text{forward} \qquad (6.119)$$

$$y''_i = \frac{1}{12h^2}(35y_i - 104y_{i-1} + 114y_{i-2} - 56y_{i-3} + 11y_{i-4}) + O(h^4) \qquad \text{backward} \qquad (6.120)$$

$$y''_i = \frac{1}{12h^2}(-y_{i+2} + 16y_{i+1} - 30y_i + 16y_{i-1} - y_{i-2}) + O(h^4) \qquad \text{centered.} \qquad (6.121)$$

Compact schemes are preferred when the order of the spatial approximation is higher than 4. Although the boundary conditions generally cannot be raised to a higher order than 4 without loss of stability, a scheme for order 6 is presented in the following section.

Compact Schemes of the Sixth Order

For simplicity, consider a mesh with uniform grid spacing, with nodes indexed by i. The generalization of Padé's method has the form [22]

$$\beta y'_{i-2} + \alpha y'_{i-1} + y'_i + \alpha y'_{i+1} + \beta y'_{i+2} = c\frac{y_{i+3} - y_{i-3}}{6h} + b\frac{y_{i+2} - y_{i-2}}{4h} + a\frac{y_{i+1} - y_{i-1}}{2h}. \qquad (6.122)$$

The relationship between coefficients a, b, c, α, and β is obtained by combining Taylor series coefficients of various orders. For $\alpha = 1/3$, $\beta = 0$,

$a = 14/9$, $b = 1/9$, and $c = 0$, one obtains the approximation of order 6 for the first-order derivative:

$$y_i' = \frac{1}{36h}(y_{i+2} + 28y_{i+1} - 28y_{i-1} - y_{i-2}) + O(h^6) \quad \text{centered.} \tag{6.123}$$

For the second-order derivative, the relationship becomes

$$\begin{aligned} \beta y''_{i-2} + \alpha y''_{i-1} + y''_i + \alpha y''_{i+1} + \beta y''_{i+2} &= c\frac{y_{i+3} - 2y_i + y_{i-3}}{9h^2} + \\ &+ b\frac{y_{i+2} - 2y_i + y_{i-2}}{4h^2} + a\frac{y_{i+1} - 2y_i + y_{i-1}}{h^2}. \end{aligned} \tag{6.124}$$

Assuming that $\alpha = 2/11$, $\beta = 0$, $c = 0$, and $b = 3/11$, yields the following approximation of the sixth order for the second-order derivative:

$$y_i'' = \frac{1}{44h^2}(3y_{i+2} + 48y_{i+1} - 102y_i + 48y_{i-1} + 3y_{i-2}) + O(h^6) \quad \text{centered.} \tag{6.125}$$

Schemes of higher order can be obtained similarly, but their application is usually restricted to particular situations, because the convergence becomes more difficult with the increase of the approximation order of the numerical algorithm. Moreover, the boundary conditions would need approximations of equivalent order, which is frequently not possible for complex geometries.

6.7 EXPLICIT AND IMPLICIT METHODS USED IN REACTIVE FLOWS

There are many numerical methods that can be employed to integrate the equations of the reactive flow. The explicit methods, which are limited in time-step due to stability conditions, are preferred when the time scale of a physical phenomenon is small. Moreover, explicit methods are of easy parallelization. Among these methods, the Jacobi and the explicit Runge-Kutta are good examples.

Among the implicit methods are the Gaussian elimination and methods such as the modified strongly implicit (MSI) procedure, the LU-SSOR, and the implicit Runge-Kutta. The parallelization of implicit methods is more elaborate than for explicit methods. Implicit methods are frequently employed for solving ill-conditioned problems, such as those that arise in reactive flows. Thus, in physical terms, implicit methods are best suited to address ill-conditioned systems, while in computational terms these methods are preferred to resolve small matrix systems.

The Gaussian elimination method corresponds to the scaling of the matrix coefficients, obtaining an upper diagonal matrix, which is solved by retrosubstitution. When the system is very ill conditioned, it is convenient to use Gaussian elimination.

6.7.1 Iterative Gauss-Seidel Method with Relaxation (Semi-Implicit)

This method is simple and can converge even when the condition of diagonal dominance of the matrix of coefficients A, for a system of type $[A]\{x\} = \{b\}$, is not satisfied. In this method, the matrix is decomposed as $[A] = [D] + [L] + [U]$, where $[D]$ corresponds to the diagonal, $[L]$ is the lower triangular matrix, and $[U]$ is the upper triangular matrix. The iterative procedure is given by

$$\{x\}^{n+1} = \frac{-[U]\{x\}^n + \{b\}}{[D] + [L]}. \tag{6.126}$$

Relaxation can be employed to accelerate the rate of convergence. It corresponds to the application of a correction to the values of a variable x as

$$x_i^{k+1} = x_i^k + w(x_i^{k+1} - x_i^k). \tag{6.127}$$

The value of relaxation parameter, w, ranges from 0 to 2, being classified as a process of over-relaxation for $1 < w < 2$, commonly used for parabolic and hyperbolic flow problems, or as a sub-relaxation, for $0 < w < 1$, which is often used in elliptic or oscillatory flows. The proper choice for the value of w, often $w \sim 2/3$ for the case of elliptical oscillatory flows and $w \sim 4/3$ for the parabolic/hyperbolic case, can reduce the computational cost needed to solve the system of discretized equations in one order of magnitude.

6.7.2 Simplified Runge-Kutta Method (Explicit)

To obtain numerical solutions of high temporal accuracy, the Runge-Kutta method can be employed [1, 2, 19]. This method is characterized by the small number of operations; more than two stages can be used to extend the stability region. Thus, the following multistage scheme, which requires less computational memory for solving $\frac{\partial \vec{W}}{\partial t} = -\vec{R}$, results in

$$\begin{aligned}
\vec{W}_{i,j,k}^{(0)} &= \vec{W}_{i,j,k}^{(n)} \\
\vec{W}_{i,j,k}^{(r)} &= \vec{W}_{i,j,k}^{(0)} - \alpha_r \Delta t \vec{R}_{i,j,k}^{(r-1)} \\
\vec{W}_{i,j,k}^{(n+1)} &= \vec{W}_{i,j,k}^{(r)},
\end{aligned} \tag{6.128}$$

where $\vec{R}_{i,j,k}^k$ corresponds to the residuum and $\vec{W} = [\bar{\rho}\tilde{u}_i \;\; \bar{\rho}\tilde{Z} \ldots \;\; \bar{\rho}\tilde{Z''}^2 \;\; \bar{\rho}\tilde{h}]$, with $k = 0, 1, 2, \ldots, n$.

The slow convergence of numerical methods is associated with attenuation of frequencies of error, $\{\varepsilon\} = \{b\} - [A]\{x\}$. The good smoothing properties of Runge-Kutta method, especially of five stages, are very important in a code developed for compressible flows. Three stages are often employed to solve incompressible flows; more steps are preferred to solve parabolic/hyperbolic flows. The coefficients for the second- and third-order temporal integration approach are given, respectively, by

1. For second order:
   three-stage: $\alpha_1 = 0.1918, \alpha_2 = 0.4929, \alpha_3 = 1.$
   five-stage: $\alpha_1 = 0.0695$, $\alpha_2 = 0.1602$, $\alpha_3 = 0.2898$, $\alpha_4 = 0.5060$, $\alpha_5 = 1.$
2. For third order:
   three-stage: $\alpha_1 = 0.2884, \alpha_2 = 0.5010, \alpha_3 = 1.$
   five-stage: $\alpha_1 = 0.1067$, $\alpha_2 = 0.1979$, $\alpha_3 = 0.3232$, $\alpha_4 = 0.5201$, $\alpha_5 = 1.$

The Runge-Kutta methods are preferred because, for a given time-step, the errors produced are smaller than those of competing methods. In addition, other properties of this method may be important to resolve stiff problems such as those that arise in reactive flows.

When using explicit methods, time-step constraints are natural; small time steps are needed to represent the mixture in a stream. In the case of compressible flows, the Fourier-Neumann condition gives for the wave equation

$$\Delta t \leq \text{CFL}\frac{\Omega}{\lambda^i + \lambda^j + \lambda^k}, \tag{6.129}$$

where CFL is the Courant-Friedrich-Lewy number, Ω is the cell volume, and λ^i is the spectral radius of the Jacobian matrix related to the direction $\vec{i}$.

An estimate for the time-varying step of the Runge-Kutta method can also be obtained from

$$h_{n+1} = h_n \left(\frac{\Delta_0}{\Delta_y}\right)^{1/(s+1)}, \tag{6.130}$$

where Δ_0 is the tolerance, s is the formal order of the method, and $\Delta_y \sim E_{n+1}$ is the instantaneous estimate of the local error.

6.7.3 Tridiagonal Matrix Algorithm (TDMA) Method

The TDMA method [9] is suitable for diagonal matrices (tridiagonal) and is of easy parallelization. Furthermore, its efficiency tends to increase with increasing mesh size. This method can be employed to directly resolve problems of the form $[A]\{x\} = \{b\}$ by writing

$$a_i x_i + b_i x_{i+1} + c_i x_{i-1} = d_i \tag{6.131}$$

or

$$x_i = P_i x_{i-1} + Q_i, \tag{6.132}$$

where a_i, b_i, and c_i are the coefficients of the matrix $[A]$ at the x_i positions, and

$$P_i = -\frac{b_i}{a_i + c_i P_{i-1}}, \tag{6.133}$$

$$Q_i = \frac{d_i - c_i Q_{i-1}}{a_i + c_i P_{i-1}}. \tag{6.134}$$

This method can be employed in each coordinate direction independently, putting the terms for other coordinate directions in d_i. So, its good properties of parallelization increase with increasing the mesh size.

6.7.4 Newton-Raphson Method (Classic)

Consider the equation $f(x) = 0$, where it finds a x_{n-1} approximation for root x. Let δ be the difference between the root x and x_{n-1}, so that

$$f(x_{n-1} + \delta) = 0. \tag{6.135}$$

Expanding the expression (6.135) in a Taylor series around x_{n-1}, yields

$$f(x_{n-1} + \delta) = f(x_{n-1}) + f'(x_{n-1})\delta + O(\delta^2) = 0. \tag{6.136}$$

If x_{n-1} is close to the root of the system and δ has a small value, so that terms of order (δ^2) can be neglected, δ can be approximated by

$$\delta \approx -\frac{f(x_{n-1})}{f'(x_{n-1})}. \tag{6.137}$$

The formula (6.136) represents the Newton-Raphson method. Using this relation, x can be determined for a new iteration according to

$$x_{n+1} = x_n - \frac{f(x_n)}{f'(x_n)}. \tag{6.138}$$

This method can be interpreted geometrically because each iteration x_{n+1} corresponds to the point where the tangent line to $f(x_{n+1})$ intersects the x-axis, as shown in Figure 6.10.

The Newton-Raphson method can present difficulties of convergence if the initial estimate for x_0 is far away from the root of the system, or if the derivative $f'(x_0)$ is close to zero.

For roots of multiplicity 1, Newton's method has quadratic convergence when x_n is close to the solution. If the multiplicity of a root is greater than 1, the convergence rate becomes linear. Knowing the multiplicity k of the root one can obtain quadratic convergence using

$$x_{n+1} = x_n - k\frac{f(x)}{f'(x)}. \tag{6.139}$$

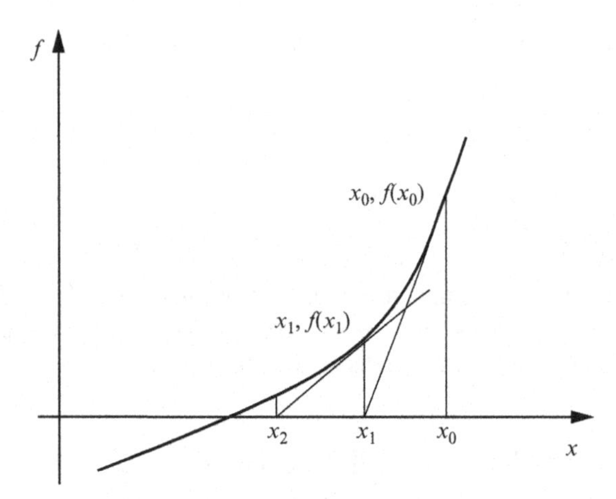

FIGURE 6.10 Graphic showing the Newton-Raphson method.

6.7.5 Modified Strongly Implicit (MSI) Procedure

In the MSI method [23], the matrix $[A]$ of the system $[A]\{x\} = \{b\}$ is decomposed as

$$[A] \sim [L][U], \tag{6.140}$$

where $[L]$ and $[U]$ are low and up triangular matrices, respectively. The product of $[L]$ and $[U]$ does not give the matrix $[A]$ identically, because some differences stored in matrix $[B]$ appear. When the convergence of the iterative process occurs, it is expected that matrix $[B]$ has values tending to zero. This allows calculation of the decomposition of the matrix $[A]$ analytically and approximately, because the inversion of large matrices is not a practical task. Thus, obtaining the numerical solution becomes an iterative process, as

$$[A+B]\{x\}^{k+1} = [A+B]\{x\}^k - ([A]\{x\}^k - \{b\}) \tag{6.141}$$

or

$$[A+B]\{x\}^{k+1} = [A+B]\{x\}^k - \{R\}, \tag{6.142}$$

where $\{R\}$ corresponds to the residue.

6.7.6 Lower-Upper Symmetric Successive Overrelaxation (LU-SSOR) Scheme

The LU-SSOR method combines the advantages of LU factorization and the Gauss-Seidel relaxation. In general, an implicit scheme for differential equations, $\frac{\partial \vec{W}}{\partial t} = -\vec{R}$, can be formulated as [24]

$$\frac{\partial \vec{W}}{\partial t} = -\left[\beta \vec{R}^{n+1} + (1-\beta)\vec{R}^n\right] \tag{6.143}$$

with $1/2 < \beta < 1$. $\vec{R}^{n+1}$ can be expanded in Taylor series as

$$\vec{R}^{n+1} = \vec{R}^n + \frac{\partial \vec{R}}{\partial \vec{W}} \Delta \vec{W} + O(\Delta t^2). \tag{6.144}$$

Then, after inserting Eq. (6.144) into Eq. (6.143) yields

$$\frac{\partial \vec{W}}{\partial t} = -\left(\beta \vec{R}^n + (1-\beta)\vec{R}^n\right) - \beta \frac{\partial \vec{R}}{\partial \vec{W}} \Delta \vec{W} \sim \frac{\Delta \vec{W}}{\Delta t} \tag{6.145}$$

or

$$\Delta \vec{W} = -\vec{R}^n \Delta t - \beta \Delta t \frac{\partial \vec{R}}{\partial \vec{W}} \Delta \vec{W}, \tag{6.146}$$

or implicitly

$$\left(1 + \beta \Delta t \frac{\partial \vec{R}}{\partial \vec{W}}\right) \Delta \vec{W} = -\vec{R}^n \Delta t. \tag{6.147}$$

This system can be decomposed as

$$(LD^{-1}U)\Delta \vec{W} = -\Delta t \vec{R}^n. \tag{6.148}$$

One iteration of the LU-SSOR scheme is done in two steps, as:

$$L\Delta \vec{W}^* = -\Delta t \vec{R}^n, \tag{6.149}$$

$$U\Delta \vec{W} = D\Delta \vec{W}^*, \tag{6.150}$$

$$\vec{W}^{n+1} = \vec{W}^n + \Delta \vec{W}. \tag{6.151}$$

The computational cost of the LU-SSOR scheme is comparable to that of the two-step explicit scheme. The damping properties of the error of the LU-SSOR method tend to be a bit worse when compared to explicit multistep methods, such as the simplified Runge-Kutta method. However, implicit or semi-implicit methods are preferred to solve stiff systems of equations.

6.8 METHODS FOR SOLVING STIFF SYSTEMS OF EQUATIONS

Stiffness can be defined as the resistance of an elastic body to deflect or distort upon application of a force. The definition in computational fluid dynamics corresponds to how hard is to solve a system of equations. The efficiency loss can be seen as a problem of stiffness, because it is a direct consequence of the flow.

The system of coupled equations for combustion has very different time scale characteristics. In geochemistry, the situation is similar, but a little less critical.

The rigidity of the system of differential equations depends on the equations, their initial conditions, and the numerical method. Nonstiff methods can be employed to solve stiff problems, but these require much more computational time. One example is the propagation of flame fronts. The stiffness of the equations has to do with complex chemical process and differences in time scales.

The exponential source term of Arrhenius equations depends strongly on the temperature, which varies considerably in many reactive flows. This variation is transferred to the density, which affects the stability of the numerical code. So, relaxation can be employed for calculating the density. In addition, the dynamic pressure variations are small but important and need to be properly evaluated to determine local effects in reactive flows.

The numerical methods that are better for solving stiff problems perform more computational work in each step and can take larger steps, such as the implicit methods. There are also solutions that were developed to solve stiff systems of equations; among the solutions applied to problems of moderate size appears the Differential Algrebraic System Solver (DASSL). In this solution, the derivatives are approximated by the backward formula, and the nonlinear system resulting in each step is solved by the Newton's method.

A decrease in the exponents of a reaction, as is often done in many reduced kinetic schemes, can lead to a reduced stiffness. In the reaction rate, the exponential term $\mathrm{e}^{-T_a/T}$ can be replaced by $\mathrm{e}^{-\beta/\alpha}\mathrm{e}^{-\beta(1-\theta)/[1-\alpha(1-\theta)]}$, where $\alpha \sim 0.75$, and $\beta \sim 8$ [13]. In general, the maximum reaction rate occurs for $\theta = (T - T_u)/(T_b - T_u)$ of order $(1 - 1/\beta)$. Subindices b and u indicate the burned and unburned conditions, respectively. Poinsot and Veynante [13] indicate that for a simple system of chemical equations, the most successful procedure is the use of chemical schemes of moderate stiffness with small β and fine meshes throughout the domain, due to transient flame movements.

The main factors that affect the time-step size are precision and stability. Precision (accuracy) is related to the magnitude of the local error while stability refers to the decrease of the error in the iterative process.

A measure of the stiffness can be given by the index

$$S = \frac{T}{J} = -T\,\mathrm{Real}(\lambda), \tag{6.152}$$

where T is the time/period of interest (e.g., one cycle), and $\mathrm{Real}(\lambda)$ is the real part of the eigenvalue λ of the Jacobian matrix with higher negative magnitude (spectral radius). It can be said that if S is of the order of magnitude 10^3, the system is stiff, and if the order of S is 10^0, the system is well conditioned. Actually, S indicates how severe the situation is [25].

For a system of the type

$$\frac{\mathrm{d}\vec{C}}{\mathrm{d}t} = \vec{f}, \tag{6.153}$$

the Jacobian matrix corresponds to the derivative $\partial\vec{f}/\partial\vec{C}$. If the negative real part of the eigenvalues of the Jacobian matrix is very large and the time cycle (period) is significant, its product indicates that the system is *stiff*, requiring special procedures to solve it.

6.8.1 Euler/Newton's Method for Stiff Systems

The backward differential formulas (BDF) have also been used to solve stiff systems of equations, with orders ranging from one to five. For example, solving

$$y_{n+1}^k = y_n^k + hf_{n+1}(t_{n+1}, y_{n+1}^k) \tag{6.154}$$

with $k = 1, 2, 3, \ldots$ (Euler method) decreases the computational cost by using a variation of Newton's method, making

$$F(y_{n+1}) = y_{n+1} - y_n - hf_{n+1} = 0 \tag{6.155}$$

or

$$y_{n+1}^k = y_n^k - \frac{F(y_{n+1}^k)}{F'(y_{n+1}^k)} \tag{6.156}$$

with

$$F'(y_{n+1}^k) = I - hbJ(t_{n+1}, y_{n+1}^k), \tag{6.157}$$

where $b = 1$ for BDF of order 1, $b = 2/3$ for BDF of second order, $b = 6/11$ for BDF of order 3, $b = 12/25$ for BDF of order 4, and $b = 60/137$ for BDF of order 5. The method becomes

$$y_{n+1}^{k+1} = y_{n+1}^k - \frac{F(y_{n+1}^k)}{I - hbJ(t_{n+1}, y_{n+1}^k)}. \tag{6.158}$$

Note that the Jacobian matrix usually varies little in each iteration, so its value does not need to be recalculated in all iterations, reducing the computational time needed for the simulation.

6.8.2 Rosenbrock Methods

Rosenbrock methods are L-stable methods based on the class of Runge-Kutta methods. Implicit or semi-implicit Runge-Kutta methods are known to satisfy the condition of strong stability.

Rosenbrock methods require the solution of a linear system of equations at each step, simplifying the solution of the problem compared to the implicit Runge-Kutta method. It is defined by

$$y_{n+1} = y_n + h\sum_{i=1}^{s} w_i k_i, \tag{6.159}$$

$$k_i = f\left(y_n + h\sum_{j=1}^{i-1} a_{ij}k_j\right) + h\,d\frac{\partial f}{\partial y}\left(y_n + \sum_{j=1}^{i-1} b_{ij}k_j\right) k_i. \tag{6.160}$$

For $w_1 = 0$, $w_2 = 1$, $a_{21} = (\sqrt{2} - 1)/2$, $b_{21} = 0$, and $d = 1 - \sqrt{2}/2$, the Rosenbrock method of order 2 becomes [26]

$$y_{n+1} = y_n + hk_2, \tag{6.161}$$

$$k_1 = f(y_n)/A(y_n), \tag{6.162}$$

$$k_2 = f(y_n + h\, a_{21}k_1)/A(y_n), \tag{6.163}$$

where

$$A(y_n) = [I - h\, d\, \partial f(y_n)/\partial y]. \tag{6.164}$$

The Rosenbrock method of order 4 results in

$$y_{n+1} = y_n + h \sum_{i=1}^{4} w_i k_i \tag{6.165}$$

$$k_1 = f(y_n)/A(y_n) \tag{6.166}$$

$$k_2 = f(y_n + h\, a_{21}\, k_1)/A(y_n) \tag{6.167}$$

$$k_3 = f(y_n + h\, (a_{31}\, k_1 + a_{32}\, k_2))/A(y_n) \tag{6.168}$$

$$k_4 = f(y_n + h\, (a_{41}\, K_1 + a_{42}\, k_2 + a_{43}\, k_3))/A(y_n), \tag{6.169}$$

whose coefficients are shown in Table 6.1.

The local estimation of the error is given by

$$E_{n+1} = \frac{\|y^*_{n+1} - y_{n+1}\|}{(2^s - 1)}, \tag{6.170}$$

where s is the order of the method and the norm $\|.\|$ is given by

$$\|y\| = \left[\frac{1}{m} \sum_{i=1}^{m} \left(\frac{y^i_{n+1}}{y^i_{\max}} \right)^2 \right]^{1/2}. \tag{6.171}$$

y^*_{n+1} is computed using time step h, while y_{n+1} is calculated using time step $h/2$. A simple estimate of the error can be obtained by the difference of the values of k_3 and k_4, reducing the cost of the computation.

Given a tolerance ϵ, the following procedure is adopted to determine the variable time-step:

1. If $E_{n+1} > \epsilon$, the step is rejected and h is reduced.
2. If $3/4\epsilon < E_{n+1} < \epsilon$, the step is accepted, but h is reduced.
3. If $\epsilon/10 < E_{n+1} < 3\epsilon/4$, the step is accepted, and h is accepted.
4. If $E_{n+1} < \epsilon/10$ the step is accepted and h is increased.

TABLE 6.1 Coefficients for the Fourth-Order Rosenbrock Method [26]

Parameter	Value
w_1	0.9451564786
w_2	0.3413231720
w_3	0.5655139575
w_4	−0.8519936081
d	0.5728160625
a_{21}	−0.5000000000
a_{31}	−0.1012236115
a_{32}	0.9762236115
a_{41}	−0.3922096763
a_{42}	0.7151140251
a_{43}	0.1430371625

To avoid the inconvenience of divergence in the iterative process, the time step may be limited by adopting

$$h_{n+1} = h_n \min\left[10, \max\left(0.1; \frac{0.9}{\Delta y}\right)^2\right], \tag{6.172}$$

considering the following restrictions:

- h values must be limited by a maximum $h_{\max}$, and a minimum $h_{\min}$.
- The initial step should be small enough.
- After a rejection of h_{n+1}, the growth factor in the next iteration step (immediately after rejection) is made equal to 1 instead of 10 in Eq. (6.172).

In the next section, the principal types of boundary conditions frequently found in reactive flow situations are discussed.

6.9 APPLICATION OF BOUNDARY CONDITIONS

In general, the following possibilities exist for defining the boundary conditions:

1. The variable value is provided at the border (Dirichlet condition).
2. The flux is provided (Neumann condition).
3. The boundary conditions are cyclic or periodic.

Obtaining proper boundary conditions for reactive flows is very important, especially when employing recent methods for solving the transient Navier-Stokes equations (DNS and LES). For such applications, most numerical schemes need high precision. The restrictions imposed on the boundary conditions for transient simulations of high order are as follows [13]:

- Transient simulations of compressible flows (DNS or LES) require precise control of the reflections of the acoustic waves at the boundary of the computational domain.
- Dissipative schemes do not require high order for transient simulations, because the acoustic waves and numerical waves propagate over long distances and times and they interact with the boundary conditions.
- Exact boundary conditions can be obtained for the Euler equations, but the problem is much more complex for the Navier-Stokes equations.
- The computational results depend on the original equations and boundary conditions.

Physically, the boundary conditions are frequently classified as of impermeable wall, cut and symmetry, far field, and periodic.

6.9.1 Permeable and Impermeable Walls

Impermeable Wall

The boundary condition for an impermeable wall without slipping (no-slip condition) states that the tangential component of velocity vector equals the speed of the wall, while the normal component has value zero. The impermeable wall imposes a zero flow condition in the normal direction. The gradients for variables such as temperature, mixture fraction, mass fraction, and pressure are set equal to zero, that is, $\partial\phi/\partial n = 0$ where $\phi = \{T, Z, Y_F, Y_{O_2}, Y_{CO_2}, Y_{H_2O}, \ldots, p\}^T$. Figure 6.11 left side illustrates a velocity profile where this type of boundary condition occurs.

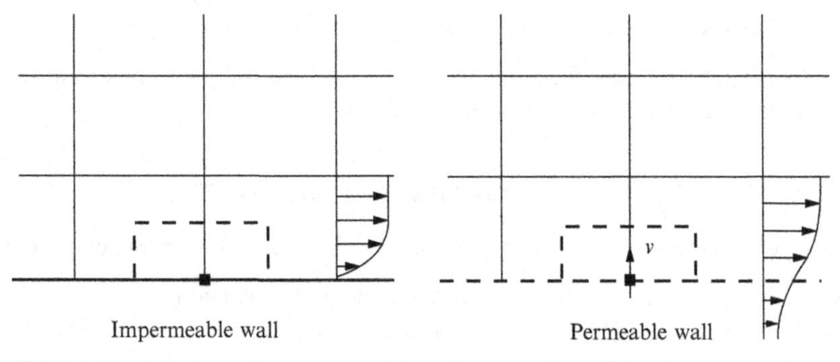

FIGURE 6.11 Velocity profile over permeable and impermeable walls.

The fluid can be considered continuous if the mean free path (distance a molecule moves between two successive collisions) is small compared to the characteristic length of the flow [27]. By analyzing the continuum, the fluid velocity along a fixed impermeable wall must be zero. Even for turbulent flows with high velocity gradient near the wall, the nonslip condition must be satisfied at every point of the wall. However, in flows where the density of the fluid is very low, as in the case of rarefied gases, there is the appearance of discontinuities and thus requires an analysis at the molecular level.

Permeable Wall

In the case of a permeable wall, there is a nonzero flow inside the wall. Figure 6.11 right side shows a possible speed profile for the flow over a permeable wall. The velocity of the flow may be determined by Darcy's law:

$$u_f = -\frac{K}{\mu}\frac{\mathrm{d}p}{\mathrm{d}x}, \tag{6.173}$$

where K is the permeability of the medium and μ the viscosity. The velocity in the permeable wall interface (u_s) differs from the average velocity within the permeable medium (u_f). Shear effects are transmitted to the porous body through the boundary layer. Considering the slip speed proportional to shear normal to the wall yields the following boundary condition:

$$\left.\frac{\mathrm{d}u}{\mathrm{d}y}\right|_{y=0} = \beta\left(u_s - u_f\right), \tag{6.174}$$

where β is a constant that depends the fluid and on the permeable media.

6.9.2 Symmetry and Cut

A cut may be employed when the flow is symmetric (for a relatively low Reynolds number). Thus, the value of velocity v for the component in the y-direction is reflected on the cut, such as,

$$v_{j=1} = -v_{j=2}, \tag{6.175}$$

as shown in Figure 6.12.

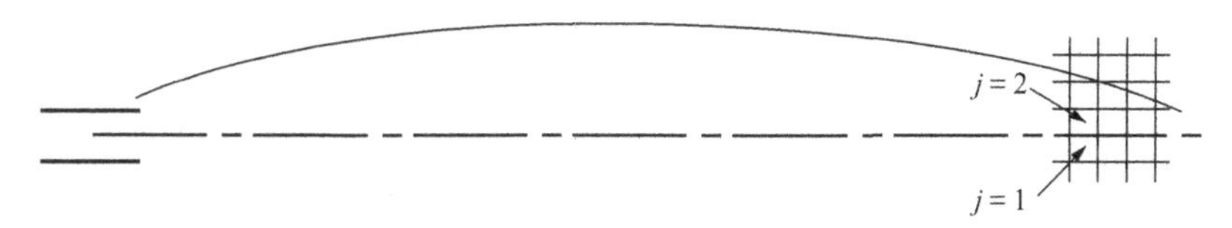

FIGURE 6.12 Condition of symmetry for laminar flow (bidimensional).

6.9.3 Far-Field

The far-field condition is obtained from the Euler equations, although the same condition can also be used with the Navier-Stokes equations. Assuming that the boundary away from the body coincides with a line y constant, the one-dimensional flow normal to the boundary is given by

$$\frac{\partial \vec{W}}{\partial t} + J\frac{\partial \vec{W}}{\partial y} = 0, \tag{6.176}$$

where $\vec{W} = \{\rho, \rho u, \rho v, \rho w, \rho H, \rho Y_F, \rho Y_{O_2}, \rho Y_{CO_2}, \rho Y_{H_2O}, \ldots, p\}^T$ and J is the Jacobian matrix, that is, $J = \frac{\partial \vec{F}}{\partial \vec{W}}$, where $\vec{F}$ represents the convective terms.

The Jacobian matrix can be diagonalized via

$$J = P \Lambda P^{-1}, \tag{6.177}$$

where the columns of P are the eigenvectors corresponding to the eigenvalues of J. Multiplying Eq. (6.176) by P^{-1}, one obtains

$$P^{-1}\frac{\partial \vec{W}}{\partial t} + P^{-1}P \Lambda P^{-1}\frac{\partial \vec{W}}{\partial y} = 0, \tag{6.178}$$

and the resultant vector $\vec{C}$ at boundary is

$$\vec{C} = P^{-1}\vec{W}. \tag{6.179}$$

The correction values on the far field boundary are obtained from the equation for $\vec{C}$ for points a and b shown in Figure 6.13, different for the fluid entrance and exit of the domain, as

$$P^{-1}\left.\vec{W}\right|_a = P^{-1}\left.\vec{W}\right|_b. \tag{6.180}$$

In the case of $\vec{W} = \{\rho, u, v, w\}^T$ along the inlet for subsonic flow (Figure 6.13 left side),

$$\rho_b = \rho_a + \frac{p_b - p_a}{c_0^2}, \tag{6.181}$$

$$u_b = u_a \pm \frac{p_a - p_b}{\rho_0 c_0}, \tag{6.182}$$

$$v_b = v_a \pm \frac{p_a - p_b}{\rho_0 c_0}, \tag{6.183}$$

$$p_b = \frac{1}{2}\{p_a + p_l \pm \rho_0 c_0[(u_a - u_l) + (v_a - v_l)]\}, \tag{6.184}$$

where $\rho_0 c_0$ represents the product of density by the reference sound speed, with $c_0 = u + c$ for the compressible case and $c_0 = \sqrt{u_0^2(1 - Ma^2)^2 + \beta^2 c^2}$,

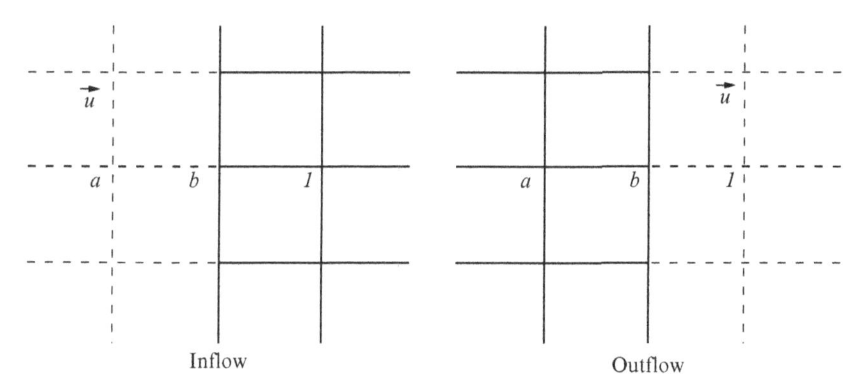

FIGURE 6.13 Boundary conditions away from the body (far field).

with $\beta \simeq 0.3$, for the incompressible case. The sign $\pm$ is associated with the increase/decrease of the mesh lines counter.

For the outflow, the result for subsonic flow is

$$p_b = p_l \tag{6.185}$$

$$\rho_b = \rho_a + \frac{p_b - p_a}{c_0^2}, \tag{6.186}$$

$$u_b = u_a \pm \frac{p_a - p_b}{\rho_0 c_0}, \tag{6.187}$$

$$v_b = v_a \pm \frac{p_a - p_b}{\rho_0 c_0}. \tag{6.188}$$

A similar outcome can be obtained in three dimensions and for $\overrightarrow{W} = \{\rho, \rho u, \rho v \rho w, \rho E, \rho Y_F, \rho Y_{O_2}, \ldots, p\}^T$, resulting in more complex relations.

The application of far field boundary condition can reduce significantly the domain needed for the solution of external flows.

6.9.4 Periodic Boundary Condition

The periodic boundary condition is applied when the boundary is repeated periodically, as shown in Figures 6.14 and 6.15.

This boundary condition is also widely adopted in large systems to model only a small part of them, away from the extremities, called a unit cell. Application of this boundary condition allows one to ignore the surface effects on the extremities of the domain. Thus, the system can be represented as a set of identical unit cells to form an infinite network as shown in Figure 6.15. In this case, the flow leaving the cell enters on its opposite side, with a rotating effect. Thus, the condition is equivalent to removing the boundaries from the system.

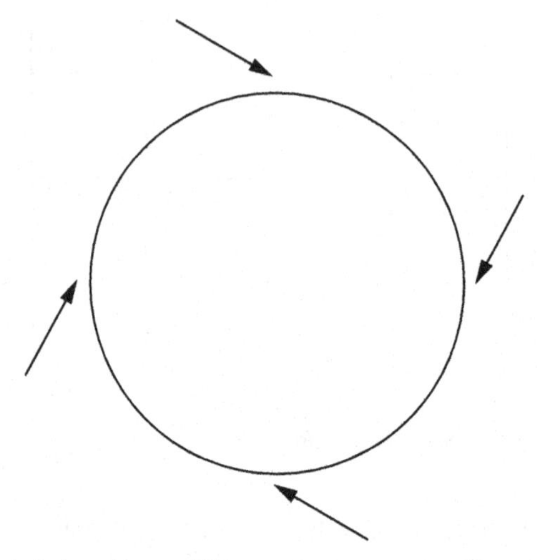

FIGURE 6.14 Periodic boundary condition.

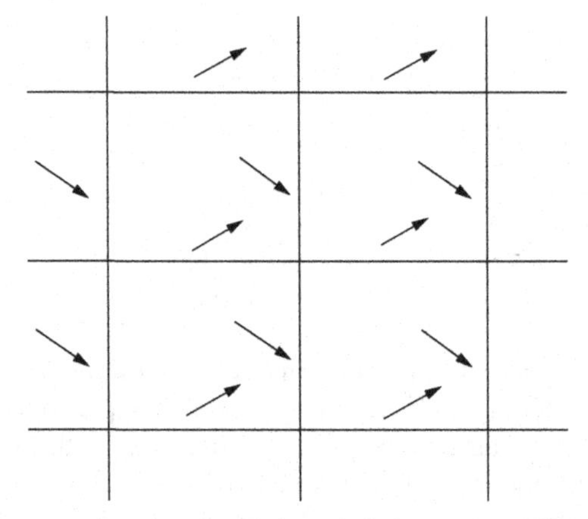

FIGURE 6.15 Representation of a unit cell of a periodic boundary condition.

6.10 SOME TECHNIQUES TO ACCELERATE THE CONVERGENCE

Most iterative methods possess the characteristic of lowering the rate of convergence after a few iterations, and the total computational work is proportional to the number of cells elevated to a number greater than 1. Thus, the development of techniques to accelerate the convergence of numerical methods is important. Among these techniques, local time-stepping, residual smoothing, and the multigrid technique are introduced. Precautions should be taken not to change the transient behavior of flows.

6.10.1 Local Time Stepping

This technique considers the application of the maximum value of Δt for each cell according to the following equation [3]

$$\Delta t_{i,j,k} = \text{CFL}\frac{\Omega_{i,j,k}}{\left(\lambda^i + \lambda^j + \lambda^k\right)_{i,j,k}}, \tag{6.189}$$

where $\Omega_{i,j,k}$ is the cell volume and λ^i, λ^j, and λ^k are the spectral radius of Jacobian matrix in each coordinate direction.

The local time-step is equivalent to preconditioning the residue in each cell. This procedure can reduce the computational time required to obtain the solution in steady state by an order of magnitude.

6.10.2 Residual Smoothing

A weighted average of residues is employed to increase the Courant-Friedrich-Lewy number of a method. The smoothing for a one-dimensional problem is given by [3]

$$\vec{R}_j^* = -\varepsilon\vec{R}_{j-1} + (1 + 2\varepsilon)\vec{R}_j - \varepsilon\vec{R}_{j+1}. \tag{6.190}$$

$\varepsilon = 0.8$ for compressible flows and $\varepsilon = 9M^2$ for incompressible flow [3], where M is the local Mach number.

6.10.3 Multigrid Techniques

The multigrid technique corresponds to a convergence acceleration method for solving differential equations by using meshes of different sizes. Multigrid methods are considered the fastest for solving elliptic partial differential equations. Moreover, they are among the fastest methods for solving several other problems with various partial differential equations, and integral equations, among others [28].

The numerical solution of a system of differential equations in very fine meshes presents slow convergence. The multigrid method transforms the low frequency error components on the fine grid to high frequency error components in a coarse mesh, which can be damped by relaxation, thereby accelerating the rate of convergence of the system.

The use of this technique allows a savings of up to 90% of the computational work required when the Runge-Kutta method is combined with the residual smoothing for compressible and incompressible flows [3].

Interpolation functions are required for the transfer of variables between meshes. An operator interpolation/prolongation maps meshes $2h$ to h. A restriction operator maps the functions of the meshes h to $2h$. The most commonly

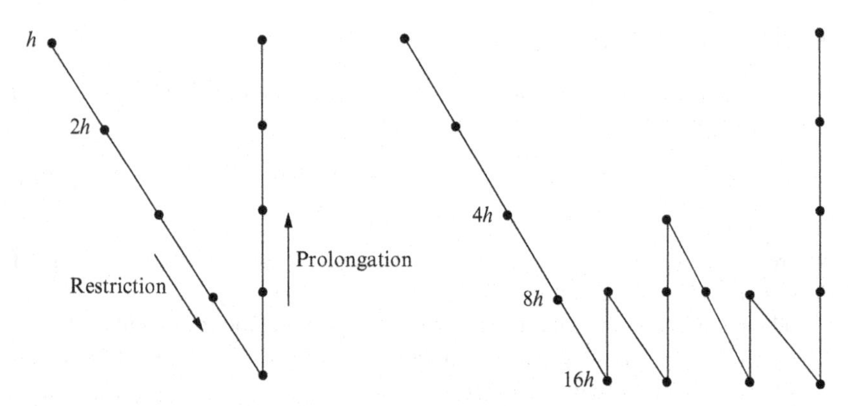

FIGURE 6.16 V and W multigrid cycles.

used interpolation method is bilinear interpolation. The restriction operator most used for this purpose is the total approximation operator (full approximation). The restriction and prolongation operators are given by, respectively:

$$I_h^{2h} = \frac{1}{4}\begin{bmatrix} 1 & 2 & 1 \\ 2 & 4 & 2 \\ 1 & 2 & 1 \end{bmatrix}, \quad I_{2h}^{h} = \frac{1}{16}\begin{bmatrix} 1 & 2 & 1 \\ 2 & 4 & 2 \\ 1 & 2 & 1 \end{bmatrix}. \tag{6.191}$$

There are basically two types of multigrid cycles: V and W, as shown in Figure 6.16. The V cycle is more appropriate for solving supersonic and hypersonic flows, while the W cycle is used more for transonic and subsonic flows. For each loop some time steps (Runge-Kutta) are executed, variables are injected in the coarse grid, the residues are restricted, and the system is solved in the coarse grid. To return to the fine mesh, interpolate the corrections of coarse meshes to fine meshes.

To illustrate the Full Approximation Storage (FAS) scheme, consider the Euler equations written for fine mesh using the finite volume method, as shown in Figure 6.17:

$$\frac{\mathrm{d}}{\mathrm{d}t}\vec{W}_h + \frac{\vec{R}_h}{V_h} = 0. \tag{6.192}$$

The numerical procedure for the V cycle with two grids consists of the following steps:

1. The solution is obtained by applying (1-2) steps for the fine mesh.

$$\vec{W}_{i,j}^{n+1} = \vec{W}_{i,j}^{(5)} \tag{6.193}$$

2. Inject the variables in the coarse grid.

$$\vec{W}_{i,j}^{(0)} = \vec{W}_{i,j}^{n+1} \tag{6.194}$$

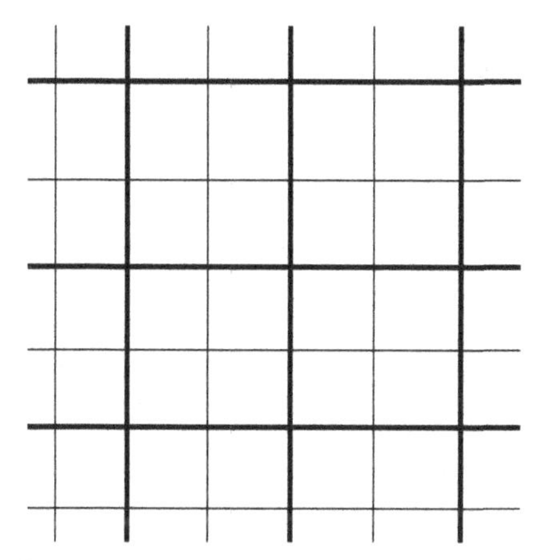

FIGURE 6.17 Fine and coarse grids.

3. Transfer the residue of fine mesh to the coarse mesh.

$$\vec{f}_{2h} = I_h^{2h} \vec{R}_h \left(\vec{W}_{i,j}^{n+1} \right) - \vec{R}_{2h}^{(0)} \tag{6.195}$$

4. Resolve the problem in the coarse grid, performing (1 and 2) steps.

$$\vec{W}_{2h}^{(k)} = \vec{W}_{2h}^{(0)} - \alpha_k \frac{\Delta t_{2h}}{V_{2h}} \left(\vec{R}_h^{(k-1)} + \vec{f}_{2h} \right) \tag{6.196}$$

$$\vec{W}_{2h}^{n+1} = \vec{W}_{2h}^{(m)} \tag{6.197}$$

5. Interpolate the correction of the coarse mesh solution to the fine mesh.

$$\vec{C}_{2h} = \vec{W}_{2h}^{n+1} - \vec{W}_{2h}^{(0)} \tag{6.198}$$

$$\vec{C}_h = \vec{W}_h^{(5)} - \vec{W}_h^{(0)} + I_h^{2h} \vec{C}_{2h} \tag{6.199}$$

6. The solution is corrected in the fine mesh.

$$\vec{W}_h^{n+1} = \vec{W}_h^{(5)} + I_{2h}^h \vec{C}_{2h} \tag{6.200}$$

For a multigrid process with more than two loops, necessary in numerical practice, steps 2-4 are repeated successively beginning with the fine mesh. Then, one applies steps 5-6 from the coarse mesh to the fine mesh. For W cycles, more work is carried out in the coarse meshes.

For low-speed flow, difficulties appear to smooth the low frequencies of error and, consequently, to obtain good convergence rate. To alleviate this problem, more work is performed in each mesh, namely two to three steps are performed for each mesh and in combination with preconditioning. In addition, the coarse mesh may also be used to obtain a good initial approximation of the fine grid solution.

The use of multigrid techniques is advantageous because the computational work per time step is reduced and because large volumes in coarse meshes can result a quick overall balance of the solution [3].

6.11 IMPLEMENTATION ISSUES AND ANALYSIS OF UNCERTAINTIES

6.11.1 Implementation Issues

Length scales in combustion frequently range from nanometers to meters, and nine orders of magnitude are needed to capture all these scales. Three-dimensional meshes of the order of one million nodes capture around three orders of magnitude. This way, at least six orders of magnitude are modeled.

Burners, for example, have lengths of the order of 1 m, the vortex set containing more energy has diameters of about 1-10 cm. The vortices, which promote the mixture, have diameters ranging from 0.1 to 10 mm. The diffusive scales have dimensions ranging from 10 to 100 μm, and molecular iterations occur in scales of the order of 1-10 nm. For flows in general, the range of scales of length and time is shown in Figure 6.18.

The time step is limited in terms of the numerical method used and in terms of discretization employed.

The length and time scales in geochemical problems are usually much larger than in combustion problems. Simulations may be done for hundreds of thousands of years in domains of the order of 10 km. Because of that, the water reactions in geochemistry are considered to be in steady state compared to the reactions that occur with minerals. The time steps may be of the order of 1 day to 10 years, or greater.

Consider, for example, an unidimensional convective problem

$$\frac{\partial u}{\partial t} + c\frac{\partial u}{\partial x} = 0. \tag{6.201}$$

The application of Fourier analysis (Von Neumann), which assumes that a numeric solution can be represented by a true solution, u_T added by an error, ϵ, results in

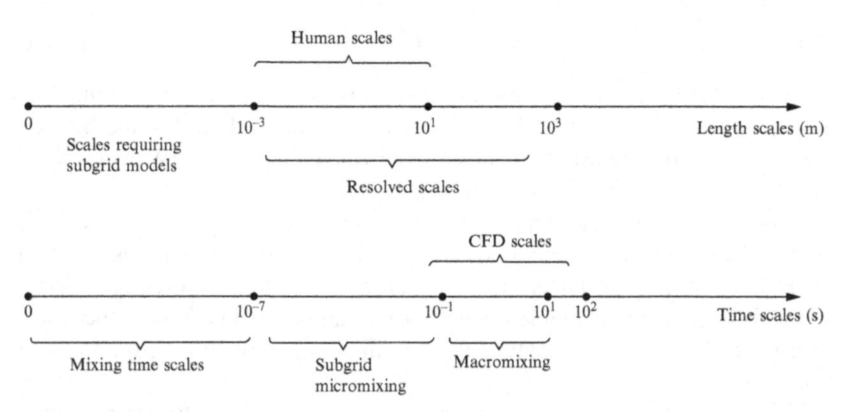

FIGURE 6.18 Representation of length and time scales.

$$u_N = u_T + \epsilon. \tag{6.202}$$

It is assumed that the error can be written as

$$\epsilon = \mathrm{e}^{at}\mathrm{e}^{ikx}, \tag{6.203}$$

and after replacing Eq. (6.203) in Eq. (6.201) one obtains

$$\frac{\partial \epsilon}{\partial t} + c\frac{\partial \epsilon}{\partial x} = 0. \tag{6.204}$$

The solution of this equation provides the stability criteria

$$\text{CFL} \sim \frac{c\Delta t}{\Delta x} \tag{6.205}$$

or

$$\Delta t \leq \frac{\text{CFL}\ \Delta x}{c}, \tag{6.206}$$

where CFL (frequently of order 10^{-1}) is the Courant-Friedrich-Lewy number, Δx the grid size, and c the characteristic speed of sound in the medium. The characteristic speed depends on the flow. In the incompressible case, the following expression is adopted:

$$c_0 \sim u(1 + M^2) + \sqrt{u^2(1 - M^2)^2 + 4M^2c^2}, \tag{6.207}$$

where M is the Mach number.

For a diffusion problem, in one dimension

$$\frac{\partial u}{\partial t} = D\frac{\partial^2 u}{\partial x^2}, \tag{6.208}$$

applying the Fourier analysis and assuming the same hypothesis for the error expansion, is obtained as a criterion

$$\text{FO} \sim \frac{D\Delta t}{\Delta x^2} \leq \frac{1}{2} \tag{6.209}$$

or

$$\Delta t \leq \frac{\text{FO}\ \Delta x^2}{D}, \tag{6.210}$$

where FO is the Fourier number.

Because Δx is small, Δx^2 is very small, and when D assumes values of the order 1, the Δt is smaller than that for the convective case.

For the chemical problem,

$$\frac{\partial Y}{\partial t} = \frac{\dot{w}}{\rho}, \tag{6.211}$$

disregarding the convective and diffusive terms, after application of Fourier analysis: the following condition results

$$\text{CHEM} \sim \frac{\Delta t\, \dot{w}}{\rho} \tag{6.212}$$

or

$$\Delta t \leq \frac{\text{CHEM}}{\dot{w}/\rho}. \tag{6.213}$$

Because $\dot{w}$ often has high values, the chemical time step is some orders of magnitude smaller than the time steps due to the effects of convection and diffusion in combustion. On the other hand, the time step may assume large values in geochemistry, because $\dot{w}$ is much smaller.

An estimate of the time step for combustion can be obtained from the ignition time step

$$t_{ig} \sim \mathrm{e}^{-E_a/RT}. \tag{6.214}$$

Because the exponent is of order 10, e^{-10} is of order 10^{-5}. As a result, the time step is chosen to be at least an order of magnitude less than this value, that is, $\Delta t \sim 10^{-6}$.

For turbulent flow, because the ratio between the integral and the Kolmogorov scales of length and time are given, respectively, by

$$\frac{L}{\eta} \sim Re^{3/4}, \tag{6.215}$$

$$\frac{t}{t_\eta} \sim Re^{1/2}, \tag{6.216}$$

Equation (6.217) gives for free turbulence (in jets) an approximation to the number of grid points

$$N_{\text{nodes}} \sim \left(\frac{L}{\eta}\right)^3 \sim Re^{9/4} \tag{6.217}$$

and Eq. (6.218) to the number of time steps

$$N_{\Delta t} \sim \frac{t}{t_\eta} \sim Re^{1/2}, \tag{6.218}$$

so that the computational time spent to obtain the numerical solution is proportional to

$$\text{CPU} \sim Re^{11/4}, \tag{6.219}$$

which limits the direct numerical simulations to low Reynolds numbers.

6.11.2 Analysis of Uncertainties

When developing a computational code, it is advisable to eliminate as many bugs as possible. Such a task can be done by selecting a set of representative cases containing accurate solutions (analytical and/or numeric). Discretization errors can be analyzed by comparing the solution for a sequence of refined meshes and varying the time step.

The mesh should be refined where the derivatives are high, meaning, close to walls and in shear layers. Solutions should be obtained with at least three meshes.

Estimation of modeling errors is a complex topic in CFD. Some turbulence models are very sensitive to the levels of the turbulence condition leading to result changes for small variations in the parameters. Even with small geometry changes, the variation in the results can be significant.

The main steps for a quantitative analysis of an error in CFD can be given by:

1. Obtaining a mesh with local refinement in regions of rapid variation of the flow and curvature of the wall.
2. Refining the mesh systematically.
3. Solving the flow with at least three meshes and comparing the solutions with reliable references to estimate the errors of the model.

Credibility of the results is obtained by proving that the results contain acceptable levels of uncertainty or error. Uncertainties can be quantified via verification techniques or validation techniques [29].

Verification

A verification process involves evaluating if the implementation is correct and does not contain errors in the input data. One can verify the quality of the solution by comparison of two solutions, one of high order and the other of low order, using the same code structure.

The systematic mesh refinement studies are the most common technique for quantifying the numerical uncertainties. The technique is simple because it does not require the modification of code and algorithm. When a mesh is refined, as a consequence, the time step is reduced and spatial discretization errors should approach zero asymptotically.

The order of convergence p varies from problem to problem. The error can be approximated by [29, 30]

$$E = f(h) - f_e = C\,h^p + O(p+1), \tag{6.220}$$

where f_e is the exact solution, $O(p+1)$ are terms of order $(p+1)$, h is the mesh spacing, and C is a constant. Neglecting the terms of higher order and applying the logarithm in Eq. (6.220) results in

$$\log(E) = \log(C) + p\,\log(h). \tag{6.221}$$

The convergence order can be obtained from the slope curve of $\log(E)$ versus $\log(h)$. When the exact solution of a problem is not known, it is necessary to obtain the solution in at least three meshes to extract the order of approximation of the solution p. Using constant refinement rate, $r = h_3/h_2 = h_2/h_1$, one obtains

$$p = \frac{\log\left(\frac{f_3 - f_2}{f_2 - f_1}\right)}{\log(r)}. \tag{6.222}$$

Often the overall accuracy of the solution is an order of magnitude less than the local order, which also occurs due to the application of boundary conditions.

The Richardson extrapolation method is a useful tool to obtain estimates of a continuous high-order value through a series of discrete values of lower order. A certain amount f can be expressed in series as

$$f = f_e + g_1 h + g_2 h^2 + g_3 h^3 + \cdots. \tag{6.223}$$

After obtaining f with two meshes, f_1 and f_2, and disregarding the terms of higher order, an estimate of the exact solution f_e is done as

$$f_e \sim f_1 + \frac{f_1 - f_2}{r^2 - 1} \tag{6.224}$$

for the second order. The Richardson extrapolation is not valid in regions of shocks and discontinuities in a flow, where frequently the order of numerical approximation used is 1. In the presence of shocks, terms of artificial dissipation are often employed, which may impair the quality of the converged solution.

For higher orders of approximation than 2, order p, yields

$$f_e \sim f_1 + \frac{f_1 - f_2}{r^p - 1}, \tag{6.225}$$

a formula to obtain f_e considered to be of the order $(p + 1)$.

Because the relative error, $\epsilon = (f_2 - f_1)/f_1$, does not take into account r and p, erroneous estimates may result when f_1 and f_e are small compared to $f_2 - f_1$. In this case, using the fractional error estimators is preferred

$$E_1 = \frac{\epsilon}{r^p - 1} \tag{6.226}$$

and

$$E_2 = \frac{\epsilon r^p}{r^p - 1}. \tag{6.227}$$

Grid convergence index (GCI)

This index can be obtained using two grid levels. However, three levels are recommended to properly estimate the order of approximation of convergence and to verify if the solution is in the range of asymptotic convergence. This range is reached when the spacing of the mesh h, for various errors E, results in a constant C, as

$$C = \frac{E_i}{h_i^p} \sim \text{constant}. \tag{6.228}$$

It is not necessary to divide by 2, the number of lines in each coordinate direction of a fine grid to obtain a coarse mesh. This common practice can put the solution outside of the range of asymptotic convergence. When the GCI value is small for the meshes adopted in the simulation, the solution is in the asymptotic range of convergence. Thus, the GCI can be seen as a measure of the percentage by which a simulated value is outside the value of asymptotic convergence. The value of the GCI to a fine mesh can be obtained from [29, 30]

$$\mathrm{GCI}_f = \frac{F_s|\epsilon|}{r^p - 1}, \tag{6.229}$$

where F_s is a safety factor. $F_s = 3$ is often adopted for comparison of results between two meshes and $F_s = 1.25$ for the comparison among three different meshes or more [31]. It can also be verified if the GCI values are in the range of asymptotic convergence by the relation

$$\frac{\mathrm{GCI}_{23}}{r^p\,\mathrm{GCI}_{12}} \sim 1. \tag{6.230}$$

However, to achieve grids in the asymptotic range is often not easy for engineering applications [31].

Similarly, the grid resolution necessary to achieve the desired (subindex d) level of accuracy can be obtained from

$$r_d = \left(\frac{\mathrm{GCI}_d}{\mathrm{GCI}_{23}}\right)^{1/p}. \tag{6.231}$$

When the GCI is not the same in all coordinate directions and for all variables, the estimate of final GCI is the sum of all GCIs, that is,

$$\mathrm{GCI} = \sum_{i=1}^{n} \mathrm{GCI}_i. \tag{6.232}$$

Validation

For the validation of a simulation, one must solve the equations that incorporate the model correctly [29, 30]. The validation includes the analysis of errors of discretization and modeling. Discretization errors can be reduced through proper allocation of mesh points, and they should be small compared with the uncertainty of the experiment. Comparison with experimental data requires that the accuracy of the experiments be known.

The simulations should demonstrate iterative convergence, and the results should demonstrate spatial and temporal convergence. So, it is possible to validate a model or a simulation; a code can only be validated for a range of applications whose experiments are known.

REFERENCES

[1] Kroll N, Rossow C-C. Foundations of numerical methods for the solution of Euler equations. Prepared for the Lecture F6.03 of the CCG. Braunschweig: DLR; 1989.

[2] Kroll N, Radespiel R, Rossow C-C. Accurate and efficient flow solvers for 3D applications on structured meshes. Lecture Series of von Karman Institute for Fluid Dynamics; 1994.

[3] De Bortoli AL. Introduction to computational fluid dynamics (in Portuguese). Ed UFRGS; 2000.

[4] Von Terzi D, Linnick M, Seidel J, Fasel H. Immersed boundary techniques for high-order methods. AIAA Paper 01-2918; 2001.

[5] Goldstein D, Adachi T, Sakata H. Modeling of flow between concentric cylinders with an external force field. In: 11th AIAA computational fluid dynamics conference, Orlando, Florida; 1993.

[6] Saiki E, Biringen S. Numerical simulation of a cylinder in uniform flow: application of a virtual boundary method. J Comput Phys 1996;123(36):450-465.

[7] Klein R. Numerics in combustion. Introduction to turbulent combustion. Belgium: Von Karman Institute for Fluid Dynamics; 1999.

[8] Rutland C, Ferziger JH. Simulation of flame-vortex interactions. Combust Flame 1991;84:343-360.

[9] Patankar SV. Numerical heat transfer and fluid flow. New York: McGraw-Hill; 1981.

[10] Chorin AJ. A numerical method for solving incompressible viscous flow problems. Math Comput 1968;22:745-762.

[11] Maliska CR. Heat transfer and computational fluid mechanics (in Portuguese). Rio de Janeiro: Livros Técnicos e Científicos Editora S.A.; 1995.

[12] Ferziger JH, Péric M. Computational methods for fluid dynamics. Berlin, Heidelberg: Springer-Verlag; 2002.

[13] Poinsot T, Veynante D. Theoretical and numerical combustion. Philadelphia, PA: R.T. Edwards, Inc.; 2001.

[14] Veynante D, Vervisch L. Turbulent combustion modeling. Prog Energy Combust Sci 2002;28:193-266.

[15] Sagaut P. Large eddy simulation for incompressible flows. Berlin: Springer; 1998.

[16] Harten A. High resolution schemes for hyperbolic conservation laws. J Comput Phys 1997;135:260-278.

[17] Gottlieb S, Shu CW. Total variation diminishing Runge-Kutta schemes. Math Comput 1998;67(221):73-85.

[18] Van Leer B. Towards the ultimate conservative difference scheme II. Monotonicity and conservation combined in a second-order scheme. J Comput Phys 1974;14(4):361-370.

[19] Jameson A, Schmidt W, Turkel E. Numerical solutions of the Euler equations by finite volume methods using Runge-Kutta timestepping schemes. In: AIAA 14th fluid and plasma dynamics conference, Palo Alto, California. AIAA Paper 81-1259; 1981.

[20] Pletcher RH, Tannehill JC, Anderson DA. Computational fluid mechanics and heat transfer. Boca Raton, FL: CRC Press, Taylor & Francis; 1997.

[21] Glasgow LA. Applied mathematics for science and engineering. Hoboken: Wiley; 2014.

[22] Lele S. Compact finite difference method with spectral like resolution. J Comput Phys 1996;103:16-42.

[23] Schneider GE, Zedan M. A modified strongly implicit procedure for the numerical solution of field problems. Numer Heat Transfer 1981;4:1-19.

[24] Blazek J. Computational fluid dynamics: principles and applications. Amsterdam: Elsevier Science Ltd.; 2006.

[25] Yu Y. Stiff problems in numerical simulation of biochemical and gene regulatory networks. Ph.D. thesis. Fudan University, Shanghai, China; 2000.
[26] Bui TD, Bui BT. Numerical methods for extremely stiff systems for ordinary differential equations. Appl Math Model 1979;3:355-358.
[27] Oosthuizen PH, Cascallen WE. Compressible fluid flows. New York: Mc-Graw Hill; 1997.
[28] Oosterlee CW, Trottenberg U, Schüller A. Multigrid. London: Academic Press; 2001.
[29] Roache PJ. Quantification of uncertainty in computational fluid dynamics. Ann Rev Fluid Mech 1997;29:123-160.
[30] Overview of CFD verification and validation. www.grc.nasa.gov/www/wind/valid/tutorial/overview.html [accessed January 12, 2015].
[31] Phillips T, Roy CJ. Evaluation of extrapolation-based discretization error and uncertainty estimators. In: 49th AIAA aerospace sciences meeting, Orlando, Florida; 2011.

Chapter 7

Elementary Applications

In this chapter, some numerical results for diffusion flames, flow in porous media, and combustion in porous media are shown. These flows were selected because they are representative of reactive flows of technical interest.

7.1 SOLUTION FOR DIFFUSION FLAMES

Figure 7.1 shows the computational domain used to analyze a jet diffusion flame of methane and methanol. This jet is surrounded by air, and the flame is stabilized by a pilot. This flame is comprised of a main jet with a mixture that is 25% fuel and 75% air. The mesh contains $199 \times 51 \times 51$ points.

There has been growing interest in the application of methanol as an alternative fuel that can be directly used in Otto engines or fuel cells [1]. Its potential as an alternative fuel is due to its favorable combustion properties, which include low emission of particulates and nitrogen oxides [2]. In addition, methanol is commonly used as an alcohol for producing biodiesel because of its low price, although other alcohols such as ethanol or isopropanol may yield a biodiesel fuel with better properties.

Figure 7.2 shows the mass fractions of H_2O, CO_2, and CO along the centerline of the burner based on the Sandia Flame D (Reynolds $\sim$ 22,400). The results are compared with those of Barlow and Frank for methane [3] and with those of Müller et al. [4] for methanol. Obtained results are in agreement for the combustion products, showing small discrepancies in the region of stoichiometric mixture fraction. The simulated maximum values (indicated by "num" in the figures) tend to overestimate the values of the experiment (indicated by "exp" in the figures).

The following set of equations is discretized, using the second-order space finite difference method, and solved:

$$\frac{\partial(\overline{\rho}\tilde{v}_i)}{\partial t} + \frac{\partial(\overline{\rho}\tilde{v}_i\tilde{u}_j)}{\partial x_j} = -\frac{\partial\overline{p}}{\partial x_i} + \frac{\partial}{\partial x_j}\left(\frac{1}{Re}\overline{\sigma}_{ij}\right) + \tilde{S}_{v_i}, \tag{7.1}$$

$$\frac{\partial(\overline{\rho}\tilde{Y}_i)}{\partial t} + \frac{\partial(\overline{\rho}\tilde{v}_j\tilde{Y}_i)}{\partial x_j} = \frac{\partial}{\partial x_j}\left(\bar{\rho}\overline{\mu}_t\frac{\partial\tilde{Y}_i}{\partial x_j}\right) \pm \tilde{\dot{w}}_i + \tilde{S}_{Y_i}, \tag{7.2}$$

Modeling and Simulation of Reactive Flows. http://dx.doi.org/10.1016/B978-0-12-802974-9.00007-6

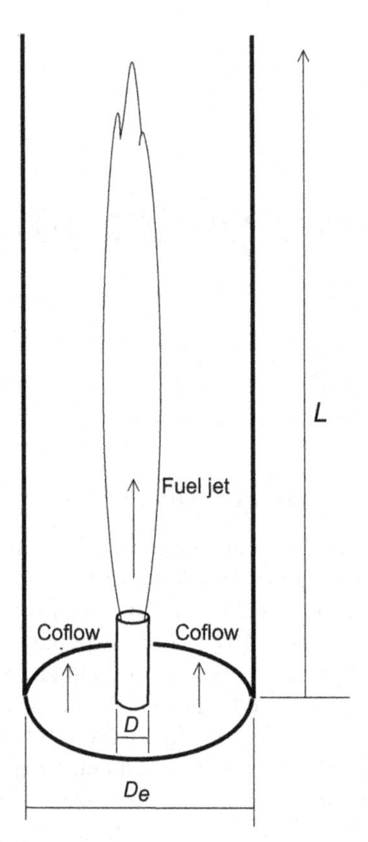

FIGURE 7.1 Sketch of the burner.

$$\frac{\partial(\bar{\rho}\tilde{h})}{\partial t} + \frac{\partial(\bar{\rho}\tilde{v}_j\tilde{h})}{\partial x_j} = \frac{\partial}{\partial x_j}\left(\bar{\rho}\bar{\mu}_t\frac{\partial\tilde{h}}{\partial x_j}\right) + \tilde{S}_h, \tag{7.3}$$

where $\bar{\rho}$ is the density, $\tilde{v}_i$ the velocity vector, $\bar{p}$ the pressure (given by a Poisson's equation),

$$\nabla^2 p = \Delta t\left(\frac{\partial\rho}{\partial t} + \vec{\nabla}\cdot(\rho\vec{v})\right), \tag{7.4}$$

where Δt is the time step. $\tilde{Y}_i$ is the mass fraction of the species $i, \tilde{h}$ the enthalpy (its relation with the temperature is given by $\tilde{h} = \sum_{i=1}^{n} \tilde{Y}_i\, h_i(\tilde{T})$), x_j the spatial coordinate, t the time, and $\tilde{T}$ the temperature. The viscous stress tensor is $\tilde{\sigma}_{ij} = (\partial\tilde{v}_i/\partial x_j + \partial\tilde{v}_j/\partial x_i) - (2/3)\delta_{ij}\partial\tilde{v}_k/\partial x_k, \delta_{ij}$ is the Kronecker delta, $\bar{\mu}_t$ the turbulent viscosity, and the reaction rate of the species i is given by $\tilde{\dot{w}}_i = W_i \sum_{k=1}^{r} \nu_{ik}\tilde{\dot{w}}_k$, where W is the molecular weight, ν the stoichiometric coefficient, and $\tilde{\dot{w}}_k$ the rate of reaction k. The turbulent viscosity comes from the Smagorinsky model.

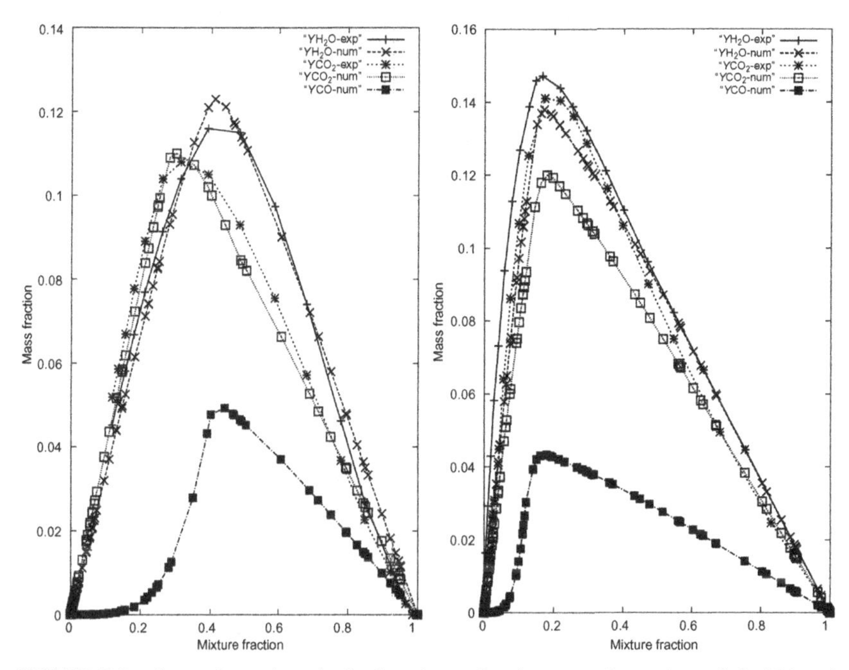

FIGURE 7.2 Comparison of results in the mixture fraction space for methane (left side) and methanol (right side).

The source terms, $\tilde{S}_{u_i}$, $\tilde{S}_h$, and $\tilde{S}_{Y_i}$, consider the overall effects of the droplets for methanol and n-heptane [5], and are given by

$$\tilde{S}_{v_i} = -\frac{1}{V}\sum_{i=1}^{N}\left(f_1\frac{m_d}{\tau_d}(u_i - u_{d,i}) + \frac{\mathrm{d}m_d}{\mathrm{d}t}u_{d,i}\right), \tag{7.5}$$

$$\tilde{S}_{Y_i} = \begin{cases} -\frac{1}{Y_{F,u}V}\sum_{i=1}^{N}\frac{\mathrm{d}m_d}{\mathrm{d}t}, & \text{se } i = F \text{ (fuel)}, \\ 0, & \text{se } i \neq F \end{cases} \tag{7.6}$$

$$\tilde{S}_h = -\frac{1}{V}\sum_{i=1}^{N}\left(\frac{1}{2\mathrm{Ec}}\frac{\mathrm{d}}{\mathrm{d}t}(m_d u_{d,i}u_{d,i}) + Q_d + \frac{\mathrm{d}m_d}{\mathrm{d}t}h_{V,S}\right), \tag{7.7}$$

where V is the cell volume, N the number of droplets, m_d the droplet mass, $\tau_d = \rho_d d_d^2/(18\mu)$ the particle response time, ρ_d the droplet density, d_d the droplet diameter, $u_{d,i}$ the droplet velocity, $Y_{F,u}$ the fuel mass fraction in the unburnt mixture, "Ec" the Eckert number, $Q_d = f_2 m_d Nu\, c_p(T - T_d)/(3Pr\, T_d)$ the heat transfer for the convection, f_1 and f_2 the corrections of the Stokes drag and heat transfer for an evaporating droplet, respectively [5], Nu the Nusselt number, $c_p = 1400\,\mathrm{J/(kg\ K)}$ the specific heat of the gas mixture, T_d the droplet temperature, and $h_{V,S}$ the enthalpy of the vapor at the droplet surface.

The Lagrangian equations for the droplets of methanol and n-heptane are given by

$$\frac{dm_d}{dt} = -\frac{\text{Sh}}{3Sc}\frac{m_d}{\tau_d}\ln(1+B_M), \tag{7.8}$$

$$\frac{du_{d,i}}{dt} = \frac{f_1}{\tau_d}(u_i - u_{d,i}) + g_i, \tag{7.9}$$

$$\frac{dT_d}{dt} = \frac{Q_d}{m_d c_{p,d}} + \frac{L_V}{m_d c_{p,d}}\frac{dm_d}{dt}, \tag{7.10}$$

where "Sh" is the Sherwood number, B_M the mass transfer number, g_i the acceleration of gravity, $c_{p,d}$ the specific heat of the liquid, $L_V = hv^0 - (c_p - c_{p,d})T_d$ the latent heat of vaporization at T_d, $hv^0 \sim 10^5$ J/K, and $c_{p,d} = 2000$ J/(kg K).

As the carbon number of the hydrocarbon fuels rises, the detailed reaction schemes become very complex. The numerical simulations of detailed kinetic mechanisms for large hydrocarbons are complicated by the existence of highly reactive radicals that induce significant stiffness to the governing equations, due to the dramatic differences in the time scales of the species. Consequently, there exists the need to develop, from these detailed mechanisms, the corresponding reduced mechanisms of fewer variables and moderated stiffness, while maintaining the accuracy and comprehensiveness of the detailed mechanism [6]. The principal chain for the n-heptane is shown in Figure 7.3.

Instantaneous maps of the mixture fraction, temperature, and main combustion products (H_2O, CO_2, CO) are shown in Figure 7.4 for the n-heptane. N-heptane is a fuel commonly used in engines. Its cetane number is approximately 56, which is typical for diesel fuel, because its properties of ignition and combustion are similar to those of diesel fuel [7]. The n-heptane has received substantial interest because it is a major component of the primary reference fuel (PRF) in internal combustion engine studies [6] and is considered a surrogate for liquid hydrocarbon fuels used in many propulsion and power generation systems [8].

Intermittent behavior is observed away from the jet entrance due to the increase of fluctuations in turbulent jets and shear layers. The high temperature regions correspond to the regions of high mass fraction of combustion products. These regions are located near the stoichiometric surface, where there are ideal conditions of burning.

The process of mixture and reaction is controlled by large vortices. The inflow is symmetric and, after the entrance, the boundary effects break down the symmetric behavior. The evolution of the mixing layer and the breakdown of the bigger vortices is verified.

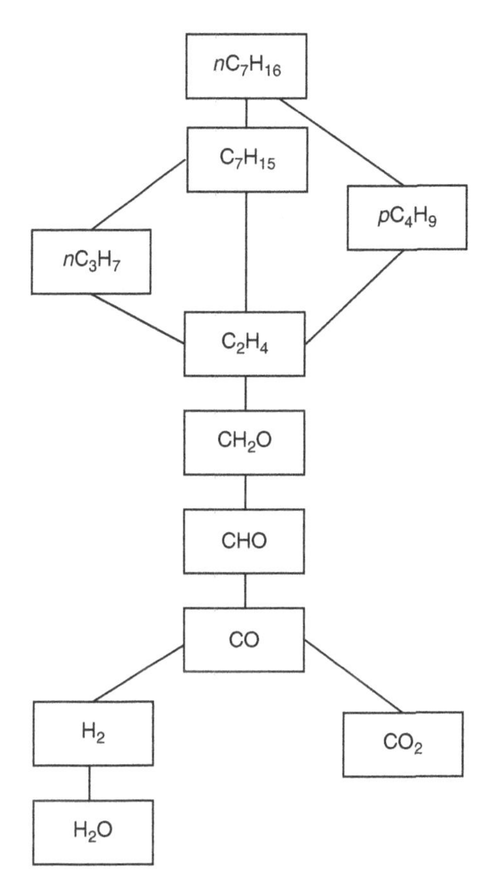

FIGURE 7.3 Principal chain for the n-heptane.

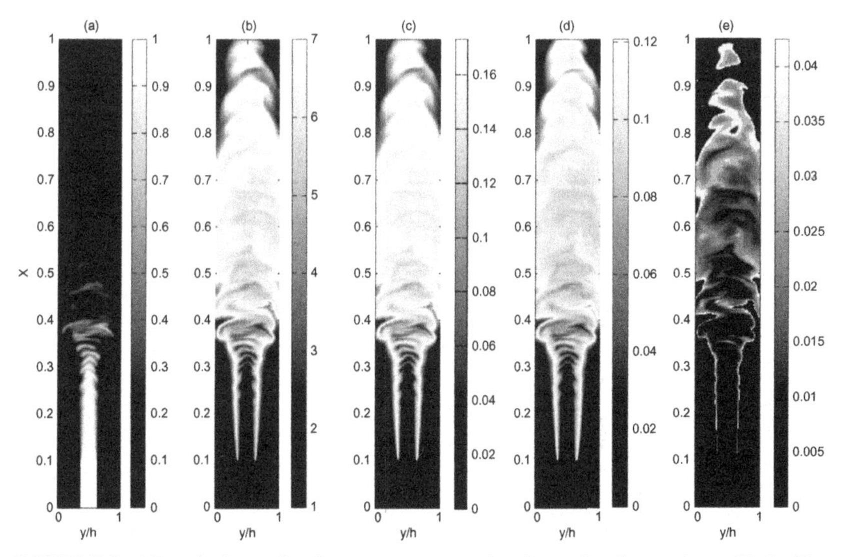

FIGURE 7.4 Map of mixture fraction, temperature, and main combustion products (H_2O, CO_2, CO) for n-heptane.

7.2 SOLUTION FOR THE FLOW IN POROUS MEDIA

Figure 7.5 shows the geometry and boundary conditions for the laminar flow in porous media. The domain is a square box of dimensions 1×1, and the computational mesh contains 51×51 points. No slip condition is adopted for velocity vector components. The other conditions, not shown in Figure 7.5, are of the Neumann type.

The complete set of equations describing the flow, heat transfer, and mass transfer phenomena with the Boussinesq approximation follows.

Momentum equation

$$\frac{\partial}{\partial t}(\phi\rho\vec{v}) + \vec{\nabla}\cdot(\phi\rho\vec{v}\vec{v}) = -\phi\vec{\nabla}p + \vec{\nabla}\cdot(\mu\phi\vec{\nabla}\vec{v}) - \frac{\mu\phi\vec{v}}{K} + \phi\vec{g}\rho\left(\beta_T(T - T_\infty) + \sum_{i=1}^{M}\beta_{C_i}(C_i - C_\infty)\right). \tag{7.11}$$

Enthalpy equation

$$\frac{\partial(\phi\rho c_P T)}{\partial t} + \vec{\nabla}\cdot(\phi\rho c_P\vec{v}T) = \vec{\nabla}\cdot(\phi\kappa_T\nabla T) - h_v(T_S - T) \pm \phi\sum_{i=1}^{N}h_i\dot{w}_i W_i. \tag{7.12}$$

Concentration of Species equation

$$\frac{\partial(\phi C_i)}{\partial t} + \vec{\nabla}\cdot(\phi\vec{v}C_i) = \vec{\nabla}\cdot(\phi D_C\vec{\nabla}C_i) \pm \phi\dot{w}_i W_i. \tag{7.13}$$

and

$$\rho = \rho_0\left[1 - \beta_T(T - T_\infty) - \sum_{i=1}^{M}\beta_{C_i}(C_i - C_\infty)\right], \tag{7.14}$$

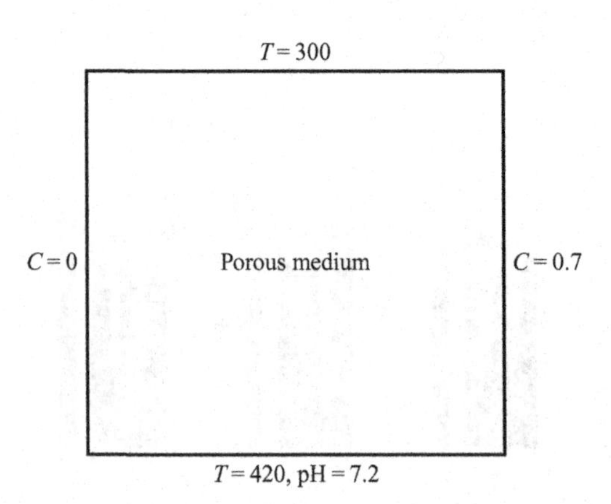

FIGURE 7.5 Boundary conditions for the flow in the square cavity.

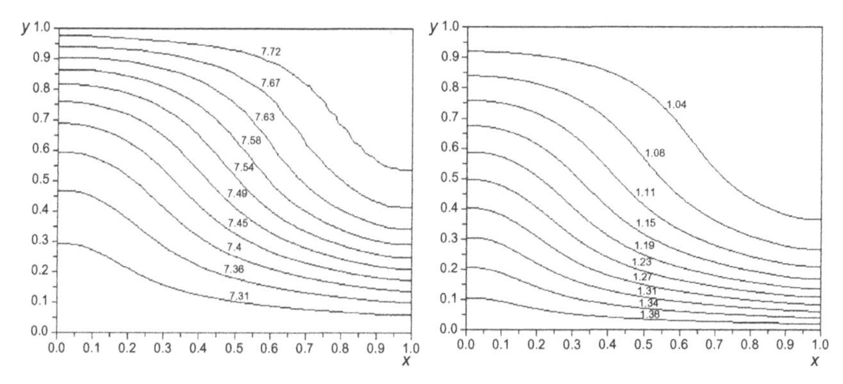

FIGURE 7.6 Isolines of pH (left side) and temperature (right side) inside a porous cavity.

where ϕ is the porosity of the medium, $\vec{v}$ the velocity vector, t the time, p the pressure, T the temperature, μ the viscosity, K the permeability, β_T and β_C the coefficients of volumetric thermal expansion and of volumetric concentration expansion, respectively, κ_T the thermal conductivity, and D_C the coefficients of mass diffusivity, C_i the concentration, and N the number of reactions. $\vec{\nabla}$ corresponds to the gradient operator, $\dot{w}_i$ is the reaction rate. A positive sign in $\pm$ represents the formation of a species, and a negative sign represents its consumption. W_i is the molecular weight of chemical species.

The following set of equations for the dissolution of calcite was solved:

$$CaCO_3 + H^+ = HCO_3{}^- + Ca^{2+} \tag{7.15}$$

$$H_2O + CO_{2(g)} = H_2CO_3{}^* \tag{7.16}$$

$$H_2CO_3{}^* = HCO_3{}^- + H^+ \tag{7.17}$$

$$HCO_3{}^- = CO_3{}^{2-} + H^+. \tag{7.18}$$

Figure 7.6 shows the pH (on the left side) and the temperature (on the right side) for the flow inside the cavity. A vortex appears in the center of the cavity due to variations of temperature and calcite concentration. The fluid moves slowly in a free convection cell. The results for the pH and temperature compare favorably with the ones from Genthon et al. [9]. While the pH (and pCO_2) decreases with the depth, the temperature increases from 1 (300 K) to 1.4 (420 k). The deflections of the lines are due to the temperature and concentration gradients. The solubility of calcite is a function of the pH of the fluid, as well as of the calcium, Ca, concentration and ions of carbonate concentration.

7.3 SOLUTION FOR COMBUSTION IN POROUS MEDIA

The combustion of methane in porous media of zirconia occurs in the domain shown in Figure 7.7. The mesh contains 90 × 51 points, and the computational

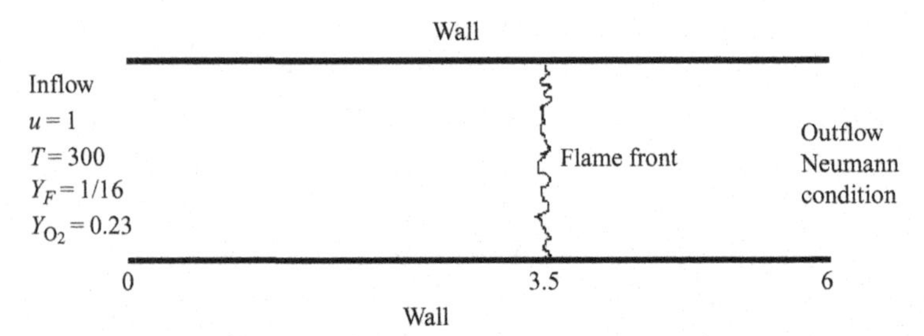

FIGURE 7.7 Geometry for combustion of methane in porous media.

domain is a rectangle of 6×1. A premixed mixture of methane-air enters the duct and, after burning, the combustion products appear.

It is assumed that the flow is steady, laminar, incompressible, and two-dimensional. The porous medium is considered to be homogeneous, isotropic, and in thermodynamic equilibrium with the saturated fluid; the gas phase radiation is neglected.

The following set of equations was solved.

Momentum

$$\frac{\partial}{\partial t}(\phi\rho\vec{v}) + \vec{\nabla}\cdot(\phi\rho\vec{v}\vec{v}) = -\phi\vec{\nabla}p + \vec{\nabla}\cdot(\phi\mu\nabla\vec{v}) + \phi\frac{\mu}{K}\vec{v}, \tag{7.19}$$

neglecting the inertia, resistance term, and fluctuation term.

Fluid energy

$$\frac{\partial}{\partial t}(\phi\rho h) + \vec{\nabla}\cdot(\phi\rho\vec{v}h) = \vec{\nabla}\cdot(\phi D_T\nabla T) - h_v(T_s - T) + \phi\sum_{i=1}^{N_{sp}}\dot{\omega}_i M_i h_i. \tag{7.20}$$

Solid matrix energy

$$\frac{\partial}{\partial t}[(1-\phi)\rho_s C_s T_s] = \vec{\nabla}\cdot[(1-\phi)D_S\nabla T_s] + h_v(T - T_s), \tag{7.21}$$

neglecting the radiation effects and the convective terms.

Chemical species

$$\frac{\partial}{\partial t}(\phi\rho Y_i) + \vec{\nabla}\cdot(\phi\rho\vec{v}Y_i) = \vec{\nabla}\cdot(\phi D\vec{\nabla}\rho Y_i) + \phi\dot{\omega}_i W_i, \tag{7.22}$$

where $\vec{v}$ is the velocity vector, ρ the fluid density, ϕ the porosity, p the pressure, K the permeability, μ the viscosity, c_p the specific heat at constant pressure, h the specific enthalpy of the gas, D_T and D_S the thermal diffusivity and the diffusivity of the solid, respectively, W_i the molecular weight of the ith species, h_v the volumetric heat transfer coefficient from the solid to the gas, T the temperature of the fluid, and T_s the temperature of the solid. The variable $Y_i = [Y_F, Y_O, Y_P]^T$ represents the mass fractions of fuel, oxygen, and products, respectively; D is the mass diffusion coefficient, and $\dot{\omega}_i$ the reaction rate.

The temperature of the fluid is obtained from the enthalpy, using a Newton iteration

$$h = \int c_p \, dT. \tag{7.23}$$

The chemical kinetics is analyzed for the combustion of methane using the following global reaction:

$$CH_4 + 2O_2 \rightarrow CO_2 + 2H_2O, \tag{7.24}$$

whose reaction rate is given by

$$\dot{\omega} = -A y_F^m y_O^n \exp\left(-\frac{\text{Ea}}{RT}\right), \tag{7.25}$$

with $A = 1.3 \times 10^8$, $\text{Ea}/R = 24.358$ K, $m = -0.3$, and $n = 1.3$ [10]. Ea is the activation energy and R is the gas constant.

Figure 7.8 shows the mass fractions of the fuel, CH_4, oxidizer O_2, and of the principal products of combustion, H_2O and CO_2. Obtained results, indicated

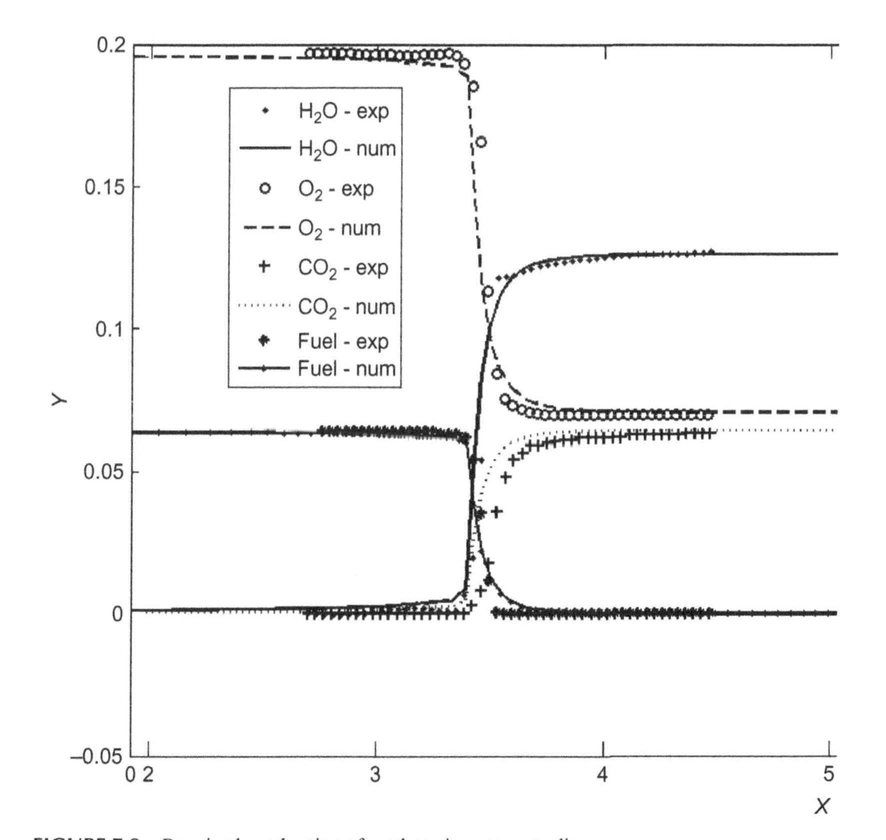

FIGURE 7.8 Premixed combustion of methane in porous media.

by "num" in the figures, compare well with the data, indicated by "exp" in the figure, presented by Barra et al. [11]. Fuel and oxidizer are consumed where the principal products of combustion appear. The mesh was refined at proximity of flame front.

REFERENCES

[1] Demirbas A. Biodiesel: a realistic fuel alternative for diesel engines. London: Springer-Verlag; 2008.

[2] Lindstedt RP, Meyer MP. A dimensionally reduced reaction mechanism for methanol oxidation. Proc Combust Inst 2002;29:1395-1402.

[3] Barlow R, Frank J. Piloted CH_4/air flames C, D, E and F—release 2.0. Sandia National Laboratories. www.ca.sandia.gov/TNF [accessed July 15, 2008].

[4] Müller CM, Seshadri K, Chen J. Reduced kinetic mechanisms for counterflow methanol diffusion flames. In: Reduced kinetic mechanisms for applications in combustion systems. Berlin, Heidelberg: Springer-Verlag; 1993.

[5] Watanabe H, Kurose R, Hwang S-M, Akamatsu F. Characteristics of flamelets in spray flames formed in a laminar counterflow. Combust Flame 2007;148:234-248.

[6] Lu T, Law CK. Strategies for mechanism reduction for large hydrocarbons: n-heptane. In: 5th US combustion meeting; 2007.

[7] Zeuch T, Moréac G, Ahmed SS, Mauss F. A comprehensive skeletal mechanism for the oxidation of n-heptane generated by chemistry guided reduction. Combust Flame 2008;155:651-674.

[8] Berta P, Aggarwal SK, Puri IK. An experimental and numerical investigation of n-heptane/air counterflow partially premixed flames and emission of NO*x* and PAH species. Combust Flame 2006;145:740-764.

[9] Genthon P, Schott J, Dandurand JL. Carbonate diagenesis during thermo-convection: application to secondary porosity generation in clastic reservoirs. Chem Geol 1997;142:41-61.

[10] Turns SR. An introduction to combustion: concepts and applications. Singapore: McGraw-Hill; 2000.

[11] Barra AJ, Diepvens G, Ellzey JL, Henneke MR. Numerical study of the effects of material properties on flame stabilization in a porous burner. Combust Flame 2003;134:369-379.

Appendix A

Review of Operations with Vectors and Tensors

This appendix describes a few basic vector and tensor operations that may be useful in understanding the material presented in the book. The vectors and tensors are presented only in the Cartesian coordinate system [1, 2].

A vector can be represented in terms of basis vectors $\vec{e}_i$ as

$$\vec{V} = V_x\vec{e}_x + V_y\vec{e}_y + V_z\vec{e}_z \tag{A.1}$$

with magnitude

$$V = |\vec{V}| = \sqrt{V_x^2 + V_y^2 + V_z^2}. \tag{A.2}$$

The scalar product of two vectors produces a scalar:

$$\vec{A} \cdot \vec{B} = A_xB_x + A_yB_y + A_zB_z, \tag{A.3}$$

and the cross product of two vectors produces a vector:

$$\vec{A} \times \vec{B} = (A_yB_z - A_zB_y)\vec{e}_x + (A_zB_x - A_xB_z)\vec{e}_y + (A_xB_y - A_yB_x)\vec{e}_z. \tag{A.4}$$

Vector differentiation has the following properties:

$$\frac{\partial}{\partial x}(\vec{A} + \vec{B}) = \frac{\partial \vec{A}}{\partial x} + \frac{\partial \vec{B}}{\partial x}, \tag{A.5}$$

$$\frac{\partial}{\partial x}(\vec{A} \cdot \vec{B}) = \vec{A} \cdot \frac{\partial \vec{B}}{\partial x} + \frac{\partial \vec{A}}{\partial x} \cdot \vec{B}, \tag{A.6}$$

$$\frac{\partial}{\partial x}(\vec{A} \times \vec{B}) = \vec{A} \times \frac{\partial \vec{B}}{\partial x} + \frac{\partial \vec{A}}{\partial x} \times \vec{B}. \tag{A.7}$$

The gradient of a scalar field produces a vector:

$$\vec{\nabla} S = \frac{\partial S}{\partial x}\vec{e}_x + \frac{\partial S}{\partial y}\vec{e}_y + \frac{\partial S}{\partial z}\vec{e}_z. \tag{A.8}$$

This formula defines a vector field in three-dimensional space called the gradient of S. At each point of a gradient field that is nonnull, the vector points in the direction where increase of S is maximal. The gradient of a vector produces a second-rank tensor [3]:

$$\vec{\nabla}\vec{V} = \begin{pmatrix} \frac{\partial V_x}{\partial x} & \frac{\partial V_y}{\partial x} & \frac{\partial V_z}{\partial x} \\ \frac{\partial V_x}{\partial y} & \frac{\partial V_y}{\partial y} & \frac{\partial V_z}{\partial y} \\ \frac{\partial V_x}{\partial z} & \frac{\partial V_y}{\partial z} & \frac{\partial V_z}{\partial z} \end{pmatrix}. \tag{A.9}$$

The curl of a vector produces a vector:

$$\vec{\nabla} \times \vec{V} = \left(\frac{\partial V_z}{\partial y} - \frac{\partial V_y}{\partial z}\right)\vec{e}_x + \left(\frac{\partial V_x}{\partial z} - \frac{\partial V_z}{\partial x}\right)\vec{e}_y + \left(\frac{\partial V_y}{\partial x} - \frac{\partial V_x}{\partial y}\right)\vec{e}_z. \tag{A.10}$$

The curl of a vector field, representing the velocity of a fluid, is related to the phenomenon of rotation of the fluid. Consider $\vec{V}$ a vector field that represents the velocity field of a fluid particle, and let a point located at (x, y, z). The particles situated in the vicinity of this point tend to rotate around the axis formed by the vector $\vec{\nabla} \times \vec{V}$.

The divergence of a vector produces a scalar:

$$\vec{\nabla} \cdot \vec{V} = \frac{\partial V_x}{\partial x} + \frac{\partial V_y}{\partial y} + \frac{\partial V_z}{\partial z}. \tag{A.11}$$

Given a vector field $\vec{V}$, such as a velocity field of a fluid, the divergence of the field is given by expansion ($\vec{\nabla} \cdot \vec{V} > 0$) or by contraction ($\vec{\nabla} \cdot \vec{V} < 0$) of the volume of gas by the flow field.

The divergence of a second-rank tensor produces a vector:

$$\begin{aligned} \vec{\nabla} \cdot \vec{\vec{\tau}} = & \left(\frac{\partial \tau_{xx}}{\partial x} + \frac{\partial \tau_{xy}}{\partial y} + \frac{\partial \tau_{xz}}{\partial z}\right)\vec{e}_x \\ & + \left(\frac{\partial \tau_{yx}}{\partial x} + \frac{\partial \tau_{yy}}{\partial y} + \frac{\partial \tau_{yz}}{\partial z}\right)\vec{e}_y \\ & + \left(\frac{\partial \tau_{zx}}{\partial x} + \frac{\partial \tau_{zy}}{\partial y} + \frac{\partial \tau_{zz}}{\partial z}\right)\vec{e}_z. \end{aligned} \tag{A.12}$$

The Laplacian of a scalar field produces a scalar:

$$\nabla^2 S = \frac{\partial^2 S}{\partial x^2} + \frac{\partial^2 S}{\partial y^2} + \frac{\partial^2 S}{\partial z^2}. \tag{A.13}$$

The Laplacian of a scalar field S is a second-order differential operator that corresponds to the divergence of the gradient of the scalar.

Physically, the Laplacian is interpreted as the concavity of the function S. Using a Taylor series development around the point (x_0, y_0, z_0), it is shown that the Laplacian at this point is proportional to the difference between the average value of the field $(\bar{S})$ to the volume element around the point and value S_0 in the field (x_0, y_0, z_0).

The Laplacian of a vector field can be written in terms of other vector operators as

$$\nabla^2 \vec{V} = \vec{\nabla}(\vec{\nabla} \cdot \vec{V}) - \vec{\nabla} \times (\vec{\nabla} \times \vec{V}), \tag{A.14}$$

which in Cartesian coordinate becomes:

$$\begin{aligned} \nabla^2 \vec{V} = & \left(\frac{\partial^2 V_x}{\partial x^2} + \frac{\partial^2 V_x}{\partial y^2} + \frac{\partial^2 V_x}{\partial z^2} \right) \vec{e}_x \\ & + \left(\frac{\partial^2 V_y}{\partial x^2} + \frac{\partial^2 V_y}{\partial y^2} + \frac{\partial^2 V_y}{\partial z^2} \right) \vec{e}_y \\ & + \left(\frac{\partial^2 V_z}{\partial x^2} + \frac{\partial^2 V_z}{\partial y^2} + \frac{\partial^2 V_z}{\partial z^2} \right) \vec{e}_z. \end{aligned} \tag{A.15}$$

Let S be a continuous differentiable scalar and vectors $\vec{V}$, $\vec{A}$, and $\vec{B}$ also continuous and differentiable. So, the following identities are valid [1, 2]:

$$\nabla^2 S = \vec{\nabla} \cdot \vec{\nabla} S, \tag{A.16}$$

$$\nabla^2 \vec{V} = (\vec{\nabla} \cdot \vec{\nabla}) \vec{V}, \tag{A.17}$$

$$\vec{\nabla} \cdot (\vec{\nabla} \times \vec{V}) = 0, \tag{A.18}$$

$$\vec{\nabla} \times \vec{\nabla} S = 0, \tag{A.19}$$

$$\vec{\nabla} \cdot (S\vec{V}) = (\vec{\nabla} S) \cdot \vec{V} + S(\vec{\nabla} \cdot \vec{V}), \tag{A.20}$$

$$\vec{\nabla} \times (S\vec{V}) = (\vec{\nabla} S) \times \vec{V} + S(\vec{\nabla} \times \vec{V}), \tag{A.21}$$

$$\vec{\nabla} \times (\vec{\nabla} \times \vec{V}) = \vec{\nabla}(\vec{\nabla} \cdot \vec{V}) - \nabla^2 \vec{V}, \tag{A.22}$$

$$(\vec{V} \cdot \vec{\nabla}) \vec{V} = \frac{1}{2} \vec{\nabla}(\vec{V} \cdot \vec{V}) - \vec{V} \times (\vec{\nabla} \times \vec{V}), \tag{A.23}$$

$$\begin{aligned} (\vec{A} \cdot \vec{\nabla}) \vec{B} = & \frac{1}{2} [\vec{\nabla}(\vec{A} \cdot \vec{B}) - \vec{\nabla} \times (\vec{A} \times \vec{B}) - \vec{B} \times (\vec{\nabla} \times \vec{A}) \\ & - \vec{A} \times (\vec{\nabla} \times \vec{B}) - \vec{B}(\vec{\nabla} \cdot \vec{A}) + \vec{A}(\vec{\nabla} \cdot \vec{B})], \end{aligned} \tag{A.24}$$

$$\vec{\nabla} \times (\vec{A} \times \vec{B}) = (\vec{B} \cdot \vec{\nabla}) \vec{A} - \vec{B}(\vec{\nabla} \cdot \vec{A}) - (\vec{A} \cdot \vec{\nabla}) \vec{B} + \vec{A}(\vec{\nabla} \cdot \vec{B}), \tag{A.25}$$

$$\vec{\nabla} \cdot (\vec{A} \times \vec{B}) = \vec{B} \cdot \vec{\nabla} \times \vec{A} - \vec{A} \cdot \vec{\nabla} \times \vec{B}, \tag{A.26}$$

$$\vec{\nabla}(\vec{A} \cdot \vec{B}) = (\vec{B} \cdot \vec{\nabla}) \vec{A} + (\vec{A} \cdot \vec{\nabla}) \vec{B} + \vec{B} \times (\vec{\nabla} \times \vec{A}) + \vec{A} \times (\vec{\nabla} \times \vec{B}). \tag{A.27}$$

The operator of the material derivative is given by

$$\frac{\mathrm{D}}{\mathrm{D}t} = \frac{\partial}{\partial t} + V_x \frac{\partial}{\partial x} + V_y \frac{\partial}{\partial y} + V_z \frac{\partial}{\partial z}. \tag{A.28}$$

In vector form, the material derivative for a vector is defined as

$$\frac{\mathrm{D}\vec{V}}{\mathrm{D}t} = \frac{\partial \vec{V}}{\partial t} + (\vec{V} \cdot \vec{\nabla}) \vec{V} = \frac{\partial \vec{V}}{\partial t} + \vec{V} \cdot (\vec{\nabla} \vec{V}), \tag{A.29}$$

where the second term can be expanded using a vector identity, such as:

$$\frac{D\vec{V}}{Dt} = \frac{\partial \vec{V}}{\partial t} + \frac{1}{2}\vec{\nabla}(\vec{V} \cdot \vec{V}) - [\vec{V} \times (\vec{\nabla} \times \vec{V})]. \tag{A.30}$$

The material derivative is the rate of change of a property over time and with the position depending on the velocity field.

Reynolds transport theorem

Let S be a closed surface limiting a region of volume V and p an intensive-extensive property related to the property P by $P = mp$ (P per unit mass). So, if $\vec{N}$ is positive normal and $\vec{V}$ the velocity vector, one obtains

$$\frac{DP}{Dt} = \int_V \frac{\partial}{\partial t} p \, dV + \int_S p \vec{V} \cdot \vec{N} \, dS. \tag{A.31}$$

Divergence theorem (Gauss's theorem or Green's theorem)

Let S be a closed surface limiting a region of volume V. Then, if $\vec{N}$ is the normal vector and $d\vec{S} = \vec{N} \cdot dS$, it follows that

$$\int_V \vec{\nabla} \cdot \vec{X} \, dV = \int_S \vec{X} \cdot \vec{N} \, dS. \tag{A.32}$$

These theorems are useful when transforming integral equations of flow in differential equations, and transforming volume integrals in surface integrals, respectively.

REFERENCES

[1] Sokolnikoff IS. Tensor analysis: theory and applications. New York: John Wiley & Sons; 1951.

[2] Jeger M, Eckmann B. Vector geometry and linear algebra for engineers and scientists. London: John Wiley & Sons; 1967.

[3] Anton H, Bivens IC, Davis S. Calculus. New York: John Wiley & Sons; 2012.

Appendix B

Equations of Fluid Dynamics in Cylindrical and Spherical Coordinates

In this appendix are presented the equations of continuity, momentum, viscous stress, energy, and binary diffusion in the rectangular, cylindrical, and spherical coordinate systems for a Newtonian fluid [1, 2].

B.1 CONTINUITY EQUATION

Vector notation

$$\frac{\partial \rho}{\partial t} + \vec{\nabla} \cdot (\rho \vec{v}) = 0. \tag{B.1}$$

Rectangular coordinates (x, y, z)

$$\frac{\partial \rho}{\partial t} + \frac{\partial}{\partial x}(\rho v_x) + \frac{\partial}{\partial y}\left(\rho v_y\right) + \frac{\partial}{\partial z}(\rho v_z) = 0. \tag{B.2}$$

Cylindrical coordinates (r, θ, z)

$$\frac{\partial \rho}{\partial t} + \frac{1}{r}\frac{\partial}{\partial r}(\rho r v_r) + \frac{1}{r}\frac{\partial}{\partial \theta}(\rho v_\theta) + \frac{\partial}{\partial z}(\rho v_z) = 0. \tag{B.3}$$

Spherical coordinates (r, θ, ϕ)

$$\frac{\partial \rho}{\partial t} + \frac{1}{r^2}\frac{\partial}{\partial r}\left(\rho r^2 v_r\right) + \frac{1}{r \sin\theta}\frac{\partial}{\partial \theta}(\rho v_\theta \sin\theta) + \frac{1}{r \sin\theta}\frac{\partial}{\partial \phi}\left(\rho v_\phi\right) = 0. \tag{B.4}$$

B.2 MOMENTUM EQUATION

Vector notation

$$\rho\left(\frac{\partial \vec{v}}{\partial t} + \vec{v} \cdot \vec{\nabla}\vec{v}\right) = -\vec{\nabla}p + \vec{\nabla} \cdot \vec{\vec{\tau}} + \rho \vec{g}. \tag{B.5}$$

Rectangular coordinates (x, y, z)
x component

$$\rho\left(\frac{\partial v_x}{\partial t}+v_x\frac{\partial v_x}{\partial x}+v_y\frac{\partial v_x}{\partial y}+v_z\frac{\partial v_x}{\partial z}\right)=-\frac{\partial p}{\partial x}+\frac{\partial \tau_{xx}}{\partial x}+\frac{\partial \tau_{yx}}{\partial y}+\frac{\partial \tau_{zx}}{\partial z}+\rho g_x. \tag{B.6}$$

y component

$$\rho\left(\frac{\partial v_y}{\partial t}+v_x\frac{\partial v_y}{\partial x}+v_y\frac{\partial v_y}{\partial y}+v_z\frac{\partial v_y}{\partial z}\right)=-\frac{\partial p}{\partial y}+\frac{\partial \tau_{xy}}{\partial x}+\frac{\partial \tau_{yy}}{\partial y}+\frac{\partial \tau_{zy}}{\partial z}+\rho g_y. \tag{B.7}$$

z component

$$\rho\left(\frac{\partial v_z}{\partial t}+v_x\frac{\partial v_z}{\partial x}+v_y\frac{\partial v_z}{\partial y}+v_z\frac{\partial v_z}{\partial z}\right)=-\frac{\partial p}{\partial z}+\frac{\partial \tau_{xz}}{\partial x}+\frac{\partial \tau_{yz}}{\partial y}+\frac{\partial \tau_{zz}}{\partial z}+\rho g_z. \tag{B.8}$$

Cylindrical coordinates (r, θ, z)
r component

$$\rho\left(\frac{\partial v_r}{\partial t}+v_r\frac{\partial v_r}{\partial r}+\frac{v_\theta}{r}\frac{\partial v_r}{\partial \theta}-\frac{v_\theta^2}{r}+v_z\frac{\partial v_r}{\partial z}\right)=-\frac{\partial p}{\partial r}+\frac{1}{r}\frac{\partial (r\tau_{rr})}{\partial r}+\frac{1}{r}\frac{\partial \tau_{r\theta}}{\partial \theta}-\frac{\tau_{\theta\theta}}{r}+\frac{\partial \tau_{rz}}{\partial z}+\rho g_r. \tag{B.9}$$

θ component

$$\rho\left(\frac{\partial v_\theta}{\partial t}+v_r\frac{\partial v_\theta}{\partial r}+\frac{v_\theta}{r}\frac{\partial v_\theta}{\partial \theta}+\frac{v_r v_\theta}{r}+v_z\frac{\partial v_\theta}{\partial z}\right)=-\frac{1}{r}\frac{\partial p}{\partial \theta}+\frac{1}{r^2}\frac{\partial (r^2\tau_{r\theta})}{\partial r}+\frac{1}{r}\frac{\partial \tau_{\theta\theta}}{\partial \theta}+\frac{\partial \tau_{\theta z}}{\partial z}+\rho g_\theta. \tag{B.10}$$

z component

$$\rho\left(\frac{\partial v_z}{\partial t}+v_r\frac{\partial v_z}{\partial r}+\frac{v_\theta}{r}\frac{\partial v_z}{\partial \theta}+v_z\frac{\partial v_z}{\partial z}\right)=-\frac{\partial p}{\partial z}+\frac{1}{r}\frac{\partial (r\tau_{rz})}{\partial r}+\frac{1}{r}\frac{\partial \tau_{\theta z}}{\partial \theta}+\frac{\partial \tau_{zz}}{\partial z}+\rho g_z. \tag{B.11}$$

Spherical coordinates (r, θ, ϕ)
r component

$$\rho\left(\frac{\partial v_r}{\partial t}+v_r\frac{\partial v_r}{\partial r}+\frac{v_\theta}{r}\frac{\partial v_r}{\partial \theta}+\frac{v_\phi}{r\sin\theta}\frac{\partial v_r}{\partial \phi}-\frac{v_\theta^2+v_\phi^2}{r}\right)=-\frac{\partial p}{\partial r}+\frac{1}{r^2}\frac{\partial (r^2\tau_{rr})}{\partial r}+\frac{1}{r\sin\theta}\frac{\partial (\tau_{r\theta}\sin\theta)}{\partial \theta}+\frac{1}{r\sin\theta}\frac{\partial \tau_{r\phi}}{\partial \phi}-\frac{\tau_{\theta\theta}+\tau_{\phi\phi}}{r}+\rho g_r. \tag{B.12}$$

θ component

$$\rho\left(\frac{\partial v_\theta}{\partial t}+v_r\frac{\partial v_\theta}{\partial r}+\frac{v_\theta}{r}\frac{\partial v_\theta}{\partial \theta}+\frac{v_\phi}{r\sin\theta}\frac{\partial v_\theta}{\partial \phi}+\frac{v_r v_\theta}{r}-\frac{v_\phi^2\cot\theta}{r}\right)=-\frac{1}{r}\frac{\partial p}{\partial \theta}$$

$$+\frac{1}{r^2}\frac{\partial(r^2\tau_{r\theta})}{\partial r}+\frac{1}{r\sin\theta}\frac{\partial(\tau_{\theta\theta}\sin\theta)}{\partial\theta}+\frac{1}{r\sin\theta}\frac{\partial\tau_{\theta\phi}}{\partial\phi}+\frac{\tau_{r\theta}}{r} \tag{B.13}$$

$$-\frac{\tau_{\phi\phi}\cot\theta}{r}+\rho g_\theta. \tag{B.14}$$

ϕ component

$$\rho\left(\frac{\partial v_\phi}{\partial t}+v_r\frac{\partial v_\phi}{\partial r}+\frac{v_\theta}{r}\frac{\partial v_\phi}{\partial \theta}+\frac{v_\phi}{r\sin\theta}\frac{\partial v_\phi}{\partial \phi}+\frac{v_\phi v_r}{r}+\frac{v_\theta v_\phi\cot\theta}{r}\right)=$$

$$-\frac{1}{r\sin\theta}\frac{\partial p}{\partial\phi}+\frac{1}{r^2}\frac{\partial(r^2\tau_{r\phi})}{\partial r}+\frac{1}{r}\frac{\partial\tau_{\theta\phi}}{\partial\theta}+\frac{1}{r\sin\theta}\frac{\partial\tau_{\phi\phi}}{\partial\phi} \tag{B.15}$$

$$+\frac{\tau_{r\phi}}{r}+\frac{2\tau_{\theta\phi}\cot\theta}{r}+\rho g_\phi. \tag{B.16}$$

B.2.1 Viscous Stress Components

Vector notation

$$\vec{\vec{\tau}}=\mu\left[\vec{\nabla}\vec{v}+\left(\vec{\nabla}\vec{v}\right)^T-\frac{2}{3}\left(\vec{\nabla}\cdot\vec{v}\right)\vec{\vec{I}}\right]. \tag{B.17}$$

Rectangular coordinates (x, y, z)

$$\tau_{xy}=\tau_{yx}=\mu\left(\frac{\partial v_x}{\partial y}+\frac{\partial v_y}{\partial x}\right) \tag{B.18}$$

$$\tau_{yz}=\tau_{zy}=\mu\left(\frac{\partial v_y}{\partial z}+\frac{\partial v_z}{\partial y}\right) \tag{B.19}$$

$$\tau_{xz}=\tau_{zx}=\mu\left(\frac{\partial v_x}{\partial z}+\frac{\partial v_z}{\partial x}\right) \tag{B.20}$$

$$\tau_{xx}=\mu\left[2\frac{\partial v_x}{\partial x}-\frac{2}{3}\left(\vec{\nabla}\vec{v}\right)\right] \tag{B.21}$$

$$\tau_{yy}=\mu\left[2\frac{\partial v_y}{\partial y}-\frac{2}{3}\left(\vec{\nabla}\vec{v}\right)\right] \tag{B.22}$$

$$\tau_{zz}=\mu\left[2\frac{\partial v_z}{\partial z}-\frac{2}{3}\left(\vec{\nabla}\vec{v}\right)\right]. \tag{B.23}$$

Cylindrical coordinates (r, θ, z)

$$\tau_{r\theta}=\tau_{\theta r}=\mu\left[r\frac{\partial}{\partial r}\left(\frac{v_\theta}{r}\right)+\frac{1}{r}\frac{\partial v_r}{\partial\theta}\right] \tag{B.24}$$

$$\tau_{\theta z} = \tau_{z\theta} = \mu \left(\frac{\partial v_\theta}{\partial z} + \frac{1}{r} \frac{\partial v_z}{\partial \theta} \right) \tag{B.25}$$

$$\tau_{rz} = \tau_{zr} = \mu \left(\frac{\partial v_z}{\partial r} + \frac{\partial v_r}{\partial z} \right) \tag{B.26}$$

$$\tau_{rr} = \mu \left[2 \frac{\partial v_r}{\partial r} - \frac{2}{3} \left(\vec{\nabla} \vec{v} \right) \right] \tag{B.27}$$

$$\tau_{\theta\theta} = \mu \left[2 \left(\frac{1}{r} \frac{\partial v_\theta}{\partial \theta} + \frac{v_r}{r} \right) - \frac{2}{3} \left(\vec{\nabla} \vec{v} \right) \right] \tag{B.28}$$

$$\tau_{zz} = \mu \left[2 \frac{\partial v_z}{\partial z} - \frac{2}{3} \left(\vec{\nabla} \vec{v} \right) \right]. \tag{B.29}$$

Spherical coordinates (r, θ, ϕ)

$$\tau_{r\theta} = \tau_{\theta r} = \mu \left[r \frac{\partial}{\partial r} \left(\frac{v_\theta}{r} \right) + \frac{1}{r} \frac{\partial v_r}{\partial \theta} \right] \tag{B.30}$$

$$\tau_{\theta\phi} = \tau_{\phi\theta} = \mu \left[\frac{\sin\theta}{r} \frac{\partial}{\partial \theta} \left(\frac{v_\phi}{\sin\theta} \right) + \frac{1}{r \sin\theta} \frac{\partial v_\theta}{\partial \phi} \right] \tag{B.31}$$

$$\tau_{r\phi} = \tau_{\phi r} = \mu \left[\frac{1}{r \sin\theta} \frac{\partial v_r}{\partial \phi} + r \frac{\partial}{\partial r} \left(\frac{v_\phi}{r} \right) \right] \tag{B.32}$$

$$\tau_{rr} = \mu \left[2 \frac{\partial v_r}{\partial r} - \frac{2}{3} \left(\vec{\nabla} \vec{v} \right) \right] \tag{B.33}$$

$$\tau_{\theta\theta} = \mu \left[2 \left(\frac{1}{r} \frac{\partial v_\theta}{\partial \theta} + \frac{v_r}{r} \right) - \frac{2}{3} \left(\vec{\nabla} \vec{v} \right) \right] \tag{B.34}$$

$$\tau_{\phi\phi} = \mu \left[2 \left(\frac{1}{r \sin\theta} \frac{\partial v_\phi}{\partial \phi} + \frac{v_r}{r} + \frac{v_\theta \cot\theta}{r} \right) - \frac{2}{3} \left(\vec{\nabla} \vec{v} \right) \right]. \tag{B.35}$$

B.3 ENERGY EQUATION

Vector notation

$$\rho c_p \frac{\mathrm{D}T}{\mathrm{D}t} = -\vec{\nabla} \cdot \vec{q} + \beta T \frac{\mathrm{D}p}{\mathrm{D}t} + \dot{q}_v + \vec{\vec{\tau}} : \left(\vec{\nabla} \vec{v} \right). \tag{B.36}$$

Rectangular coordinates (x, y, z)

$$\begin{aligned} \rho c_p \left(\frac{\partial T}{\partial t} + v_x \frac{\partial T}{\partial x} + v_y \frac{\partial T}{\partial y} + v_z \frac{\partial T}{\partial z} \right) &= -\left(\frac{\partial q_x}{\partial x} + \frac{\partial q_y}{\partial y} + \frac{\partial q_z}{\partial z} \right) \\ + \beta T \left(\frac{\partial p}{\partial t} + v_x \frac{\partial p}{\partial x} + v_y \frac{\partial p}{\partial y} + v_z \frac{\partial p}{\partial z} \right) &+ \dot{q}_v + \left(\tau_{xx} \frac{\partial v_x}{\partial x} + \tau_{yx} \frac{\partial v_x}{\partial y} + \tau_{zx} \frac{\partial v_x}{\partial z} \right) \\ + \left(\tau_{xy} \frac{\partial v_y}{\partial x} + \tau_{yy} \frac{\partial v_y}{\partial y} + \tau_{zy} \frac{\partial v_y}{\partial z} \right) &+ \left(\tau_{xz} \frac{\partial v_z}{\partial x} + \tau_{yz} \frac{\partial v_z}{\partial y} + \tau_{zz} \frac{\partial v_z}{\partial z} \right). \end{aligned} \tag{B.37}$$

Cylindrical coordinates (r, θ, z)

$$\rho c_p \left(\frac{\partial T}{\partial t} + v_r \frac{\partial T}{\partial r} + \frac{v_\theta}{r} \frac{\partial T}{\partial \theta} + v_z \frac{\partial T}{\partial z} \right) = - \left(\frac{1}{r} \frac{\partial (r q_r)}{\partial r} + \frac{1}{r} \frac{\partial q_\theta}{\partial \theta} + \frac{\partial q_z}{\partial z} \right)$$
$$+ \beta T \left(\frac{\partial p}{\partial t} + v_r \frac{\partial p}{\partial r} + \frac{v_\theta}{r} \frac{\partial p}{\partial \theta} + v_z \frac{\partial p}{\partial z} \right) + \dot{q}_v + \tau_{rr} \frac{\partial v_r}{\partial r} + \tau_{\theta\theta} \frac{1}{r} \left(\frac{\partial v_\theta}{\partial \theta} + v_r \right)$$
$$+ \tau_{zz} \frac{\partial v_z}{\partial z} + \tau_{r\theta} \left(r \frac{\partial}{\partial r} \left(\frac{v_\theta}{r} \right) + \frac{1}{r} \frac{\partial v_r}{\partial \theta} \right) + \tau_{rz} \left(\frac{\partial v_z}{\partial r} + \frac{\partial v_r}{\partial z} \right). \tag{B.38}$$

Spherical coordinates (r, θ, ϕ)

$$\rho c_p \left(\frac{\partial T}{\partial t} + v_r \frac{\partial T}{\partial r} + \frac{v_\theta}{r} \frac{\partial T}{\partial \theta} + \frac{v_\phi}{r \sin\theta} \frac{\partial T}{\partial \phi} \right) =$$
$$- \left(\frac{1}{r^2} \frac{\partial (r^2 q_r)}{\partial r} + \frac{1}{r \sin\theta} \frac{\partial (q_\theta \sin\theta)}{\partial \theta} + \frac{1}{r \sin\theta} \frac{\partial q_\phi}{\partial \phi} \right)$$
$$+ \beta T \left(\frac{\partial p}{\partial t} + v_r \frac{\partial p}{\partial r} + \frac{v_\theta}{r} \frac{\partial p}{\partial \theta} + \frac{v_\phi}{r \sin\theta} \frac{\partial p}{\partial \phi} \right) \tag{B.39}$$
$$+ \dot{q}_v + \tau_{rr} \frac{\partial v_r}{\partial r} + \tau_{\theta\theta} \left(\frac{1}{r} \frac{\partial v_\theta}{\partial \theta} + \frac{v_r}{r} \right) + \tau_{\phi\phi} \left(\frac{1}{r \sin\theta} \frac{\partial v_\phi}{\partial \phi} + \frac{v_r}{r} + \frac{v_\theta \cot\theta}{r} \right)$$
$$+ \tau_{r\theta} \left(\frac{\partial v_\theta}{\partial r} + \frac{1}{r} \frac{\partial v_r}{\partial \theta} - \frac{v_\theta}{r} \right) + \tau_{r\phi} \left(\frac{\partial v_\phi}{\partial r} + \frac{1}{r \sin\theta} \frac{\partial v_r}{\partial \phi} - \frac{v_\phi}{r} \right)$$
$$+ \tau_{\theta\phi} \left(\frac{1}{r} \frac{\partial v_\phi}{\partial \theta} + \frac{1}{r \sin\theta} \frac{\partial v_\theta}{\partial \phi} - \frac{\cot\theta}{r} v_\phi \right).$$

B.4 SPECIES EQUATION

Vector notation

$$\rho \frac{\mathrm{D} Y_i}{\mathrm{D} t} = - \vec{\nabla} \cdot \vec{j}_i + \dot{w}_i, \qquad i = 1, 2, \ldots, n. \tag{B.40}$$

Rectangular coordinates (x, y, z)

$$\rho \left(\frac{\partial Y_i}{\partial t} + v_x \frac{\partial Y_i}{\partial x} + v_y \frac{\partial Y_i}{\partial y} + v_z \frac{\partial Y_i}{\partial z} \right) = - \left(\frac{\partial j_{i,x}}{\partial x} + \frac{\partial j_{i,y}}{\partial y} + \frac{\partial j_{i,z}}{\partial z} \right) + \dot{w}_i. \tag{B.41}$$

Cylindrical coordinates (r, θ, z)

$$\rho \left(\frac{\partial Y_i}{\partial t} + v_r \frac{\partial Y_i}{\partial r} + \frac{v_\theta}{r} \frac{\partial Y_i}{\partial \theta} + v_z \frac{\partial Y_i}{\partial z} \right) = - \left(\frac{1}{r} \frac{\partial (r j_{i,r})}{\partial r} + \frac{1}{r} \frac{\partial j_{i,\theta}}{\partial \theta} + \frac{\partial j_{i,z}}{\partial z} \right) + \dot{w}_i. \tag{B.42}$$

Spherical coordinates (r, θ, ϕ)

$$\rho\left(\frac{\partial Y_i}{\partial t} + v_r \frac{\partial Y_i}{\partial r} + \frac{v_\theta}{r}\frac{\partial Y_i}{\partial \theta} + \frac{v_\phi}{r\sin\theta}\frac{\partial Y_i}{\partial \phi}\right) = -\left(\frac{1}{r^2}\frac{\partial (r^2 j_{i,r})}{\partial r} + \frac{1}{r\sin\theta}\frac{\partial (j_{i,\theta}\sin\theta)}{\partial \theta} + \frac{1}{r\sin\theta}\frac{\partial j_{i,\phi}}{\partial \phi}\right) + \dot{w}_i. \tag{B.43}$$

REFERENCES

[1] Aris R. Vectors, tensors and the basic equations of fluid mechanics. New York: Dover Publications; 1990.

[2] Bird RB, Stewart WE, Lightfoot EN. Transport phenomena; New York: John Wiley & Sons; 2006.

Index

I

J

K

L

M

N

O

P

R

S

T

U

V

W

Z

Printed in the United States
By Bookmasters